T0212258

# Multipath Effects in GPS Receivers

# Synthesis Lectures on Communications

Editor
**William Tranter,** *Virginia Tech*

Fundamentals of Spread Spectrum Modulation
Rodger E. Ziemer
2007

Code Division Multiple Access(CDMA)
R. Michael Buehrer
2006

Game Theory for Wireless Engineers
Allen B. MacKenzie and Luiz A. DaSilva
2006

© Springer Nature Switzerland AG 2022

Reprint of original edition © Morgan & Claypool 2016

Multipath Effects in GPS Receivers

Steven Miller, Xue Zhang, and Andreas Spanias

ISBN: 978-3-031-00554-1    paperback
ISBN: 978-3-031-01682-0      ebook

DOI 10.1007/978-3-031-01682-0

A Publication in the Springer series
*SYNTHESIS LECTURES ON COMMUNICATIONS*

Lecture #11
Series Editor: William Tranter, *Virginia Tech*
Series ISSN
Print 1932-1244   Electronic 1932-1708

# Multipath Effects in GPS Receivers

Steven Miller, Xue Zhang, and Andreas Spanias
SenSIP Center, Arizona State University

*SYNTHESIS LECTURES ON COMMUNICATIONS #11*

## ABSTRACT

Autonomous vehicles use global navigation satellite systems (GNSS) to provide a position within a few centimeters of truth. Centimeter positioning requires accurate measurement of each satellite's direct path propagation time. Multipath corrupts the propagation time estimate by creating a time-varying bias. A GNSS receiver model is developed and the effects of multipath are investigated. MATLAB™ code is provided to enable readers to run simple GNSS receiver simulations. More specifically, GNSS signal models are presented and multipath mitigation techniques are described for various multipath conditions. Appendices are included in the booklet to derive some of the basics on early minus late code synchronization methods. Details on the numerically controlled oscillator and its properties are also given in the appendix.

## KEYWORDS

GPS, multipath, antenna

# Contents

# List of Symbols

# CHAPTER 1

# Introduction

A Staten Island ferry struck a pier as it was docking on January 9th, 2013, injuring 57 people. This was not the first accident for the ferry. On October 15, 2003, a similar accident killed 11 people [1]. This disaster may have been avoided if the ferry was autonomously controlled or a vehicle trajectory warning system was installed. Global Navigation Satellite Systems (GNSS) provide the enabling technology for real-time, autonomous vehicle navigation and control in such diverse applications as construction, mining, farming, and fishing. Consequently, there is commercial interest to increase measurement accuracy and integrity while simultaneously reducing the system cost. Modern GNSS receivers provide real-time position accuracy of a few centimeters. All GNSS receivers estimate a satellite's signal time of arrival to solve for position and time. The quality of these time-of-arrival (TOA) estimates are directly dependent upon accurate tracking of the direct sequence spread spectrum (DSSS) code and carrier phase. Unfortunately, *multipath is a dominant error source* within these systems since it corrupts the signal phase estimates with a time-varying bias.

The first GNSS system was designed by the U.S. and consists of over 24 satellites at a height of 20,000 Km and an orbital period of about 12 hours [2]. The satellites are positioned such that at least four are in view from any position on the Earth with possibly greater than 10 visible depending upon the receiver's location.

## 1.1 GNSS FUNDAMENTALS AND MOTIVATING MULTIPATH MITIGATION

GNSS is fundamentally a time estimation problem utilizing signal carrier and code phase estimates. The receiver calculates the signal propagation time for each satellite in view by extracting the signal origination time and satellite orbital position from the satellite message. An over-constrained linear equation is solved to minimize the receiver position error with respect to all visible satellites. The four variables of the linear equation are receiver position, $(x, y, z)$, and local time, $t$. When greater than four satellites are visible, additional unknowns, such as atmospheric effects, can be estimated.

The requirement to estimate the local time is illustrated with a 2D position estimation example. Consider a ship navigating through a channel with three time-synchronized beacons positioned on shore. Each beacon simultaneously transmits a message that contains a time-stamp and its location. The ship can determine its position, as illustrated in Fig. 1.1, by calculating the propagation time from each beacon to the ship. The ship is located at the intersection of the three measurement circles since it is the only solution that is consistent with the distance observations, P1, P2, and P3. Now consider the timing uncertainty of the user local time. This creates a position uncertainty around each beacon as illustrated by the signal propagation rings in Fig. 1.2. Now the solution does not consist of a single point, but a volume of possible points.

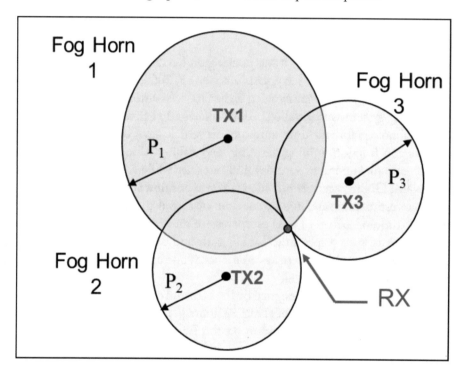

**Figure 1.1:** 2D range estimation.

Unlike the beacon example, the GNSS system employs satellites that transmit direct sequence spread spectrum (DSSS) signals over multiple L-band carriers. Transmission over multiple L-Band carriers enables receivers to measure the propagation delay through the dispersive Ionosphere for more precise positioning. Due to the carefully constructed DSSS signal and message structures, the time estimation problem becomes a code and carrier phase estimation problem where the performance of the code and carrier tracking loops determine the phase estimate resolution and ultimately the position and time resolution. Reasonable performance for modern tracking loops sets the code and carrier phase estimates at the sub-meter and centimeter level respectively. Phase measurements from all visible GNSS signals are simultaneously captured by the receiver. This enables the removal of common mode clock errors and fixed delays due to the antenna cable, down-converter analog group delay, and digital signal processing delays.

Each satellite transmits carefully constructed messages that include the satellite's position, GNSS system time, and other system parameters. Since each bit of the message perfectly aligns

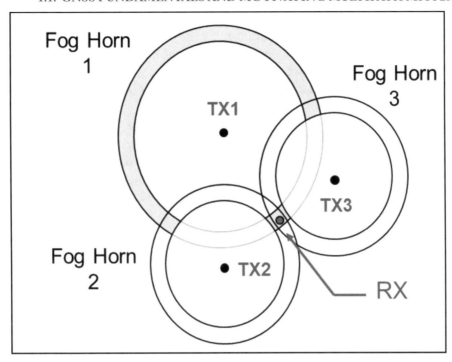

**Figure 1.2:** 2D range estimation with timing ambiguity.

with a DSSS code chip, and the receiver counts whole and fractional code chips, an absolute, unambiguous time reference is established. Therefore, the receiver can calculate the time of flight from when the satellite transmitted a certain code chip phase and when it was received. Unlike the code phase measurement which provides absolute time measurements to the transmitting satellite, the carrier phase measurement is ambiguous. Although the receiver counts the number of whole and fractional carrier cycles (modulo one phase measurement period), there is no datum for carrier cycles. Therefore, the receiver does not absolutely measure the number of whole carrier cycles occurring during the message time-of-flight; rather it measures the number of carrier cycles that accumulate between measurement observations. This measurement is referred to as the delta carrier phase, and the uncertainty in whole carrier cycles to the transmitting satellite is referred to as carrier phase ambiguity. Advanced GNSS receivers utilize the code phase estimates to limit the search for the carrier phase ambiguity. Consequently, code and carrier tracking errors, such as from multipath, degrades the phase measurement data, and therefore, the time of arrival estimate and the user location estimate.

The pseudo-range from the $i$-th satellite to the receiver is non-linear and given by,

$$\widehat{P}_i = \overline{(x_i - x_{rx})^2 + (y_i - y_{rx})^2 + (z_i - z_{rx})^2} + ct_{BIAS} + \varepsilon_i, \tag{1.1}$$

where, $x_i$, $y_i$, and $z_i$ are the positions of the $i$-th satellite given in Earth Centered Earth Fixed (ECEF) co-ordinates; $x_{rx}$, $y_{rx}$, and $z_{rx}$ are the current predicted receiver position in ECEF co-ordinates; $c$ is the speed of light; $t_{BIAS}$ is the satellite and receiver clock bias with respect to the GNSS system time; and $\varepsilon$ are systemic errors which will be described in more detail below [2]. Equation (1.1) is called the *pseudo-range* since it includes the true geometric range between the receiver and the satellite as well as error terms.

Many methods to solve Eq. (1.1) exist [2], and all suffer from multipath. An incremental approach will be presented to further introduce the fundamental concepts. Equation (1.1) is linearized by taking the first term of a Taylor series expansion to yield,

$$
\mathbf{H} = \begin{bmatrix}
\dfrac{\partial p_1}{\partial x} & \dfrac{\partial p_1}{\partial y} & \dfrac{\partial p_1}{\partial z} & 1 \\[2mm]
\dfrac{\partial p_2}{\partial x} & \dfrac{\partial p_2}{\partial y} & \dfrac{\partial p_2}{\partial z} & 1 \\[2mm]
\vdots & \vdots & \vdots & \vdots \\[2mm]
\dfrac{\partial p_N}{\partial x} & \dfrac{\partial p_N}{\partial y} & \dfrac{\partial p_N}{\partial z} & 1
\end{bmatrix},
\tag{1.2}
$$

where the partials are the unit vectors in the direction of the $N$ space vehicles in view. This matrix is also referred to the directional cosine matrix and shows how the solution depends on the geometry of the visible satellites with respect to the receiver. The receiver solves for position and time utilizing the following steps.

**Step 1:** Initially predict the receiver state, $\mathbf{X}$, given by,

$$
\mathbf{X} = \begin{bmatrix} \hat{x}_{rx} & \hat{y}_{rx} & \hat{z}_{rx} & c\hat{t}_{BIAS} \end{bmatrix}.
\tag{1.3}
$$

**Step 2:** Compute predicted pseudo ranges for all satellites in view, $\hat{p}_{i\ldots N}$. This is based upon the valid assumption that the satellites are far away and the directional cosines, $\mathbf{H}$, are the same for both the true receiver position and the current estimated position.

**Step 3:** Obtain the pseudo range from the code and carrier tracking loops, $p_{i\ldots N}$. Update the receiver state by minimizing the error between the predicted and measured pseudo range,

$$
\underbrace{\begin{bmatrix} P_1 \\ P_2 \\ \vdots \\ P_N \end{bmatrix}}_{\mathbf{Z}} - \begin{bmatrix} \hat{P}_1 \\ \hat{P}_2 \\ \vdots \\ \hat{P}_N \end{bmatrix} = \underbrace{\begin{bmatrix}
\dfrac{\partial P_1}{\partial x} & \dfrac{\partial P_1}{\partial y} & \dfrac{\partial P_1}{\partial z} & 1 \\[2mm]
\dfrac{\partial P_2}{\partial x} & \dfrac{\partial P_2}{\partial y} & \dfrac{\partial P_2}{\partial z} & 1 \\[2mm]
\vdots & \vdots & \vdots & \vdots \\[2mm]
\dfrac{\partial P_N}{\partial x} & \dfrac{\partial P_N}{\partial y} & \dfrac{\partial P_N}{\partial z} & 1
\end{bmatrix}}_{\mathbf{H}} \underbrace{\begin{bmatrix} \Delta\hat{x}_{rx} \\ \Delta\hat{y}_{rx} \\ \Delta\hat{z}_{rx} \\ c\hat{t}_{BIAS} \end{bmatrix}}_{\Delta\mathbf{X}}.
\tag{1.4}
$$

If solved by least squares, then,

$$\mathbf{Z} = \mathbf{H}\Delta\mathbf{x},$$

$$\Delta\mathbf{x} = \left(\mathbf{H}^T\mathbf{W}\mathbf{H}\right)^{-1}\mathbf{H}^T\mathbf{W}\mathbf{Z}, \quad \text{and,} \tag{1.5}$$

$$\widehat{\mathbf{x}}_{\mathbf{k}+1} = \widehat{\mathbf{x}}_{\mathbf{k}} + \Delta\mathbf{x},$$

where $\mathbf{W}$ is a weight matrix with each element corresponding to the confidence of each satellite observation. Typically, $\mathbf{W}$ is adaptive and is, at a minimum, a function of the satellite elevation angle: The lower elevation angle satellites are de-weighted since they have a lower signal-to-noise (SNR) and typically suffer a greater multipath bias, and consequently, a larger observation variance.

Recall that Eq. (1.2) implies the position solution depends on the geometry of the satellites with respect to the receiver. When the number of satellites is 4, the solution to Eq. (1.4) becomes $\Delta\mathbf{x} = \mathbf{H}^{-1}\mathbf{Z}$ and the error covariance is

$$cov\left(\Delta\mathbf{x}\right) = E\left[\Delta\mathbf{x}\Delta\mathbf{x}^T\right] = \mathbf{H}^{-1}\underbrace{E\left[\mathbf{Z}\mathbf{Z}^T\right]}_{\sigma_{RX}^2}\mathbf{H}^{-T},$$

$$cov\left(\Delta\mathbf{x}\right) = \sigma_{RX}^2\underbrace{\mathbf{H}^{-1}\mathbf{H}^{-T}}_{DOP}, \tag{1.6}$$

where $\sigma_{RX}^2$ is the measurement variance. The $\mathbf{H}^{-1}\mathbf{H}^{-T}$ term directly scales the measurement variance and is a function of the directional cosines. This term is called the Dilution of Precision (DOP) with diagonal elements consisting of the East, North, Vertical, and Time dilution of precision terms respectively. It is desirable for receivers to simultaneously utilize multiple GNSS constellations to provide a richer set of satellite geometry to decrease the DOP.

Recall that the GNSS time estimation problem was transformed to a phase estimation problem. The performance of the code and carrier tracking loops determine the phase estimate resolution and ultimately the position resolution. The measured phase range from the $i$-th satellite to a receiver, $rx$, is given by

$$P_{i,rx} = R_{i,rx} + \varepsilon_{Orbit,i} + \varepsilon_{Iono,i,rx} + \varepsilon_{Tropo,i,rx} + c\left(\varepsilon_{t,i} - \varepsilon_{t,rx}\right) \\ + \varepsilon_{MP,i,rx} + \varepsilon_{\sigma,i,rx}, \tag{1.7}$$

where

$P_{i,rx}$ = the measured phase range from the $i$-th satellite to the receiver, $rx$;

$R_{i,rx}$ = the true range from the $i$-th satellite to the receiver, $rx$;

$\varepsilon_{Orbit,i}$ = the $i$-th satellite orbital error;

$\varepsilon_{Iono,i,rx}$ = the $i$-th satellite signal propagation delay through the ionosphere;

$\varepsilon_{Tropo,i,rx}$ = the $i$-th satellite signal propagation delay through the troposphere;

$\varepsilon_{t,i}$ = the $i$-th satellite clock error (seconds) with respect to the GNSS time datum;

$\varepsilon_{t,rx}$ = the receiver clock error (seconds) with respect to the GNSS time datum;

$c$ = the speed of light in meters/second;

$\varepsilon_{MP,i,rx}$ = the code phase error due to multipath from the $i$-th satellite to the receiver; and

$\varepsilon_{\sigma,i,rx}$ = Additive White Gaussian Noise (AWGN) and other un-modeled errors.

Note that if not otherwise stated, all units are in meters.

The methods to mitigate these errors drive receiver technology, cost, and complexity. A *Pareto* chart of the error contributions is shown in Fig. 1.3. The error terms include: (1) multipath; (2) ionosphere and troposphere atmospheric errors; (3) satellite orbital errors; and (4) the receiver and satellite clock bias with respect to the GNSS system time. Each contribution is discussed below.

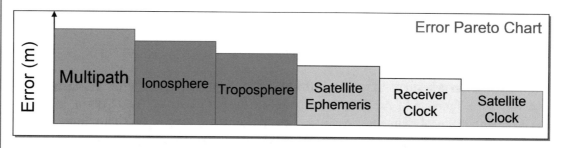

Figure 1.3: Relative positional error contributions.

Multipath is the vector sum of additional non-line-of-sight signal paths that cause a time-varying bias. Although multipath inherently has a longer time-of-flight, the resulting satellite range error can be positive or negative. GNSS transmitted signals utilize nearly time-shift orthogonal code sequences, so the code multipath error upper bound is set by the chip period: Tens of meters to hundreds of meters. Multipath is primarily mitigated by digital signal tracking algorithms and receiver antenna designs as discussed further.

Ionosphere and troposphere errors are a consequence of the satellite signals propagating through the atmosphere and are independent of the signal structure. These time-varying biases are dependent upon both the satellite and receiver positions and can be tens of meters [3]. The *ionosphere* contains charged particles that cause a frequency-dependent signal propagation bias dependent upon the sunspot activity and the Earth's crust. GNSS systems exploit the dispersive nature of the Ionosphere by transmitting on multiple L-Band frequencies to enable receivers to measure and mitigate the Ionosphere delays. The *troposphere* extends from the ground to an altitude of about 10 Km and contains the Earth's weather layer. Troposphere errors are dependent upon water vapor and altitude and are frequency independent. Differential receiver techniques are typically used to mitigate atmospheric errors.

GNSS satellites are typically not geostationary, so their transmitted messages contain their orbital position. Earth's gravitational tides and atmospheric drag can cause errors between the stated trajectory and the true satellite trajectory by about 0.5 m. These *ephemeris* errors will continuously decrease with advanced ground tracking techniques.

Satellite time is based upon atomic clocks; therefore, they are stable and accurate and any time error can be modeled as receiver errors. The stability of the local receiver time is determined by its reference clock which drifts with time, temperature, and vibration. Temperature compensated oscillators (TCXO) create a sub-meter time-varying range bias. These and other errors that are common to *all* GNSS measurements are estimated within $t_{BIAS}$ as shown in Eq. (1.1).

Commercial GNSS receivers are primarily of two types: (1) Stand alone; or (2) Differential Real-Time Kinematic (RTK). The stand-alone receiver provides real-time positioning of about 0.5 m and is the dominant type due to simplicity and cost. As the name implies, it does not require additional support equipment. In contrast, differential receivers must not only process signals from the GNSS satellites but must also receive additional signals from a base-station. Differential receivers provide centimeter level real-time positioning by removing correlated errors between the base-station and receiver but are more complex and higher cost. We shall show that multipath is the dominant error source for both types of receivers.

## 1.2    STAND-ALONE CARRIER SMOOTHED CODE RANGE MEASUREMENT MODEL

The variance of the position estimate is reduced by smoothing the code phase estimate with the carrier phase estimate. The technique, referred to as carrier-smoothing, combines two measurements: The unambiguous, but higher variance, code phase measurement and the ambiguous, but lower variance, carrier phase measurement. This method, also referred to as the Hatch Filter [3], achieves decimeter-level position resolution. It formulates the problem as a weighted one-pole IIR filter given by

$$P_k = \alpha_k P_{k-1} + (1 - \alpha_k)\left(P_{k-1} + c_k - c_{k-1}\right),  \tag{1.8}$$

where $k$ is the iteration index; $\alpha_k$ is the time varying IIR weight; $c_k$ and $p_k$ are the carrier and code phase measurements at time $k$, respectively. $P_k$ is the smoothed pseudo-range that is applied to Eq. (1.1). The weight, $\alpha_k$, is also given by an IIR structure that initializes the value near 0.99. This initially heavily weighs the code measurements to drive the solution to the unambiguous range mean contained within the noisy code measurement. The weight is gradually decreased until it saturates near 0.01 which heavily weighs the lower variance delta carrier phase measurements, thus yielding smoothed range measurements. Note that a code phase bias caused by multipath will still exist within the smoothed range, $P_k$.

## 1.3   DIFFERENTIAL RECEIVERS

A differential system consists of a base station and a rover as shown in Fig. 1.4. The base station consists of a GNSS receiver and a radio transmitter. The base station broadcasts its code and carrier phase measurements along with its location, which is typically surveyed and known to within a few millimeters. The user's receiver, called a rover, uses differencing techniques to remove base-station and rover correlated errors such as clock, ephemeris, and atmospheric errors, without explicitly estimating each error contribution. This is analogous to how a differential amplifier removes common mode errors. Consequently, errors that are time or space correlated are removed. Since base and rover multipath biases are independent, differential techniques tend to worsen the multipath bias. The inability of a differential system to mitigate multipath is the primary motivation to develop advanced multipath mitigating processing techniques.

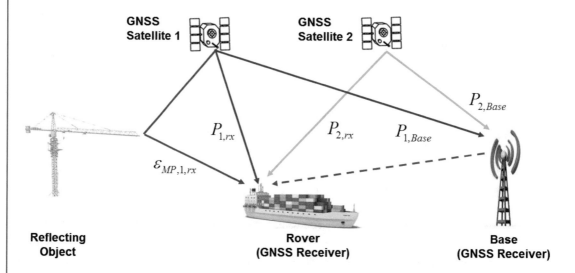

**Figure 1.4:** Base and rover differential scheme to mitigate position error.

Within Eq. (1.7), replace the subscript $RX$ with either a $B$ or an $R$ to denote a base or rover receiver respectively. Consider only satellites that are visible by both the base and the rover and assume the base and rover are close enough that they experience the same atmospheric conditions (spatially correlated atmospheric errors), then,

$$\varepsilon_{atmos,i} = \varepsilon_{Iono,i,B} + \varepsilon_{Tropo,i,B} = \varepsilon_{Iono,i,R} + \varepsilon_{Tropo,i,R}. \tag{1.9}$$

The single difference is calculated to subtract out common mode errors. The single difference is

$$P_{i,\Delta RB} = P_{i,R} - P_{i,B}. \tag{1.10}$$

Substituting Eq. (1.7) into Eq. (1.10) yields

$$P_{i,\Delta RB} = (R_{i,R} - R_{i,B})$$
$$+ (\varepsilon_{Orbit,i} + \varepsilon_{atmos,i} + c\,(\varepsilon_{t,i} - \varepsilon_{t,R}) + \varepsilon_{MP,i,R} + \varepsilon_{\sigma,i,R})$$
$$- (\varepsilon_{Orbit,i} + \varepsilon_{atmos,i} + c\,(\varepsilon_{t,i} - \varepsilon_{t,B}) + \varepsilon_{MP,i,B} + \varepsilon_{\sigma,i,B}),$$
(1.11)

which simplifies to

$$P_{i,\Delta RB} = (R_{i,R} - R_{i,B}) + (c\varepsilon_{t,R} + \varepsilon_{MP,i,R} + \varepsilon_{\sigma,i,R})$$
$$- (c\varepsilon_{t,B} + \varepsilon_{MP,i,B} + \varepsilon_{\sigma,i,B}).$$
(1.12)

Note that the atmospheric and satellite errors are removed. The remaining errors are multipath, base, and rover clock errors, as well as AWGN. An additional difference is performed again to remove the receiver clock errors. This double difference is taken between different satellites, designated as 1 and 2, which yields,

$$P_{\Delta RB} = (P_{1,R} - P_{1,B}) - (P_{2,R} - P_{2,B}).$$
(1.13)

Substituting Eq. (1.12) into Eq. (1.13) yields

$$P_{\Delta RB} = \underbrace{\{(R_{1,R} - R_{1,B}) - (R_{2,R} - R_{2,B})\}}_{Delta\,Range}$$
$$+ \underbrace{\{(\varepsilon_{MP,1,R} - \varepsilon_{MP,1,B}) - (\varepsilon_{MP,2,R} - \varepsilon_{MP,2B})\}}_{Unmitigated\,Multipath}$$
$$+ \underbrace{\{(\varepsilon_{\sigma,1,R} - \varepsilon_{\sigma,1,B}) - (\varepsilon_{\sigma,2,R} - \varepsilon_{\sigma,2,B})\}}_{noise}.$$
(1.14)

Note that the remaining errors are multipath and noise.

Base and rover correlated errors, such as atmospheric errors, satellite clock errors, and ephemeris errors, are removed by the differencing operation. However, since the base and rover are spatially separated, their respective multipath profiles are independent, and therefore the dominating error remains multipath. This is exactly what motivates advanced multipath mitigation. Multipath mitigation techniques require both digital processing techniques and an advanced antenna design.

Recent work on multipath mitigation was performed in [27–31]. We also note that research in this area has been extended to cover asymmetric correlation kernels for GPS multipath [32]. Additional work from this group is also given by [33–35].

The rest of the book is organized as follows. Chapter 2 discusses GNSS signal models. Signal structures, channel models, and receiver structures are provided. Chapter 3 reviews the existing mitigation techniques, which includes Fixed Radiation Pattern Antennas (FRPA) and signal processing methods. In Chapter 4, concluding remarks for the book are provided.

CHAPTER 2

# GNSS Signal Models

GNSS has now evolved to include multiple systems: (1) GPS operated by the U.S. [4]; (2) GLONASS developed by Russia [5]; (3) Beidou in development by China; and (4) Galileo in development by the European Union [6]. Additional regional satellite-based augmentation systems (SBAS) also exist and are operated by the United States, European Union, Japan, India, and China. The SBAS signals are similar to the GPS signal structure.

Consumer receivers utilize a limited set of the available signals; however, precision systems, such as for survey and vehicle control, adopt a "greedy" approach and utilize all of the signals available. This "greedy" approach provides: (1) More observables to reduce estimation variance in a least squares sense; (2) better geometric observability to reduce position variance due to DOP; (3) more observable frequencies for Ionsospheric and Tropospheric observability and subsequent removal; and (4) more observable frequencies to increase the RTK carrier "ambiguity wavelength" thereby reducing the number of required searchable discrete points for RTK ambiguity resolution. Moreover, to compete in some national markets, a receiver must support the corresponding national GNSS system. Table 2.1 summarizes the planned and available commercial GNSS signals [4–6].

Many satellites, frequencies, and modulation schemes exist. Commercial grade receivers will focus on carriers near 1572.450 MHz and 1176.450 MHz since a dual frequency receiver can process GPS, Galileo, and Beidou signals: L1CA, L1C, L5IQ, E1bOS, E1cOS, E5aIQ, B1, and B2IQ. Precision GNSS receivers will process all of the available signals; and consequently, a multi-path mitigation scheme must be suitable for all of the signaling types such as BPSK, QPSK, and Binary Offset Carry (BOC). The characteristics of these signaling types are described in detail below.

The Russian system utilizes frequency division multiplexing (FDMA). This has the disadvantage that all satellite signals experience a different group delay (due to the different frequency as a consequence of channelization) and therefore this complicates the positioning calculation due to this non-common mode code phase bias. It is anticipated that future Russian systems will utilize carrier division multiple access (CDMA). It should be noted that the Russian system is the only other globally deployed and fully functioning GNSS system beyond GPS, and differentially processing GLONASS signals improve receiver performance.

This chapter develops mathematical models for the transmitted satellite signals, the propagation channel, and receiver.

**Table 2.1:** Commercial GNSS signal characteristics by system [4–6]

| System | Carrier (MHz) | Transmit Filter BW (MHz) | Signal Name | Modulation | Chip Rate (Mcps) | Code Length (Chips) |
|---|---|---|---|---|---|---|
| GPS (USA) | L1: 1575.420 | 30 [4] | L1CA | BPSK | 1.023 | 1023 |
| | | | L1P(Y) | BPSK | 10.23 | 7 days |
| | | | L1C | TMBOC(m,n) | 10.23 | 10230 |
| | L2: 1227.60 | 30 [4] | L2CM-Data | TDM BPSK | 0.5115 | 10230 |
| | | | L2CL-Pilot | | 0.5115 | 767250 |
| | | | L2P(Y) | BPSK | 10.23 | 7 days |
| | L5: 1176.450 | 30 [4] | L5I-Data | QPSK | 10.23 | 10230 |
| | | | L5Q-Pilot | QPSK | 10.23 | 10230 |
| Galileo (European Union) | E1: 1575.42 | [6] | E1bOS | BOC(1,1) | 1.023 | 4092 |
| | | | E1cOS | BOC(1,1) | 1.023 | 4092 * 25 |
| | E5ab: 1191.795 | [6] | Composite | AltBOC(15,10) | 10.23 | 10230 |
| | E5a: 1176.450 | | E5a-I-Data | BPSK | 10.23 | 10230 |
| | | | E5a-Q-Pilot | BPSK | 10.23 | 10230 |
| | E5b: 1207.140 | | E5b-I-Data | BPSK | 10.23 | 10230 |
| | | | E5b-Q-Pilot | BPSK | 10.23 | 10230 |
| GLONASS (Russia) | G1-Band $f(n) = nf_o + 1602.0$ $f_o = 0.562$ $n = \{-7..+13\}$ | [5] | G1K | BPSK | 0.511 | 511 |
| | G2-Band $f(n) = nf_o + 1246.0$ $f_o = 0.4375$ $n = \{-7..+13\}$ | [5] | G2K | BPSK | 0.511 | 511 |
| Beidou (China) | B1: 1561.1 | Not Published | B1 | BPSK | 1.023 | 1023 |
| | B2: 1176.450 | Not Published | B2 | QPSK | 10.23 | 10230 |
| | B3: 1207.1 | | B3 | QPSK | | |

## 2.1   SIGNAL STRUCTURES

GNSS signals are specifically designed to facilitate code-phase estimation and can be grouped into four distinct phase-shift keyed signal structures:

1. Rectangular pulse weighted signals such as GPS L1CA;

2. Binary-offset-carrier (BOC) pulse weighted such as Galileo E1bOS;

3. Rectangular pulse weight with time interleaved data and pilot channels such as GPS L2C; and

4. Alt-BOC signals like Galileo E5ab.

The characteristics that influence signal tracking, such as energy spectral density and autocorrelation, are investigated for each signal type. The baseband signal has the form:

$$x(t) = f(t) \quad g(t) \sum_{k=-} C_n \delta(t - kT_c), \tag{2.1}$$

where $C_n$ denotes the $n$-th spreading chip with magnitudes $\{+1, -1\}$ and period $T_c$; $f(t)$ represents the impulse response of the band limited transmit filters; and $g(t)$ represents the pulse waveform of a single chip. The signal-in-space (SIS) specification for each system provides $f(t)$. For power constrained receivers, the receiver band-width is narrower than the transmit band-width. Figure 2.1 plots $g(t)$ for $Rect(n)$, $SinBOC(m,n)$, and $CosBOC(m,n)$ pulse waveforms. $Rect(n)$ describes a rectangular pulse at rate $f_c = nf_0$; with $f_0$    1.023 MHz. $BOC(m,n)$ describes a $Rect(n)$ signal further modulated by a square wave with fundamental frequency, $f_s = mf_0$, where $m$ is the number of square wave cycles within one chip period, subject to $m$, $n$, and $M = \frac{2m}{n}$ are positive integers [7].

For a signal defined with a rectangular pulse, $Rect(n)$, then $g(t)$ has the form

$$g(t) = \overline{f_c} Rect_{T_c}(t) = \begin{cases} \overline{f_c}, & \frac{-T_c}{2} \quad t \quad \frac{T_c}{2} \\ 0, & \text{Otherwise,} \end{cases} \tag{2.2}$$

and the power spectrum

$$S_{RECT}(f)^2 = T_c \sin c^2(\pi T_c f) = T_c \left( \frac{\sin(\pi T_c f)}{\pi T_c f} \right)^2. \tag{2.3}$$

For a signal defined with a $SinBOC(m,n)$ pulse, $g(t)$ has the form

$$g(t) = \begin{cases} \overline{f_c} \, sign(\sin(2\pi m f_c t)), & \frac{-T_c}{2} \quad t \quad \frac{T_c}{2} \\ 0, & \text{Otherwise,} \end{cases} \tag{2.4}$$

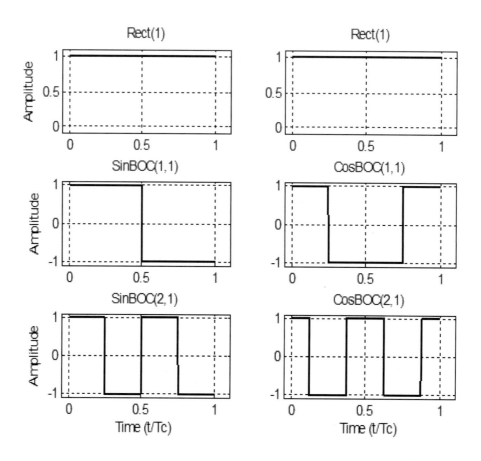

**Figure 2.1:** $g(t)$ waveforms for $Rect(n)$, $SinBOC(m, n)$, and $CosBOC(m, n)$.

which is equivalent to

$$
g(t) = \begin{cases} \overline{f_c} \sum_{k=0}^{m} (-1)^k Rect_{T_s} (t - kT_s), & \dfrac{-T_c}{2} \; t \; \dfrac{T_c}{2} \\ 0, & \text{Otherwise,} \end{cases} \tag{2.5}
$$

where $T_s = \frac{T_c}{2m}$ is half the square wave period for the secondary code. The corresponding power spectrum is [7]:

$$
S_{SBOC}(f) = \left| \begin{array}{ll} 4T_c \sin c^2 \left( \pi f T_c \right) \left[ \dfrac{\sin^2 \left( \dfrac{\pi f}{4 f_s} \right)}{\cos \left( \dfrac{\pi f}{2 f_s} \right)} \right]^2 , & k \text{ even} \\[3em] 4T_c \dfrac{\cos^2 \left( \pi f T_c \right)}{\left( \pi f T_c \right)^2} \left[ \dfrac{\sin^2 \left( \dfrac{\pi f}{4 f_s} \right)}{\cos \left( \dfrac{\pi f}{2 f_s} \right)} \right]^2 , & k \text{ odd}, \end{array} \right. \tag{2.6}
$$

with $k = \frac{2m}{n} = \frac{2 f_s}{f_c} = \frac{T_c}{T_s}$; again with $T_s$ half the square wave period.

Similarly for $CosBOC(m, n)$ signals, $g(t)$ have the form,

$$
g(t) = f_c \sum_{i=0}^{2m-1} \sum_{k=0}^{1} (-1)^{i+k} Rect_{T_s} \left( t - i T_s - \frac{k T_S}{2} \right). \tag{2.7}
$$

The power spectrum of $CBOC(m, n)$ is given by [7]:

$$
S_{CBOC}(f) = \left| \begin{array}{ll} 4T_c \sin c^2 \left( \pi f T_c \right) \left[ \dfrac{\sin^2 \left( \dfrac{\pi f}{4 f_s} \right)}{\cos \left( \dfrac{\pi f}{2 f_s} \right)} \right]^2 , & k \text{ even} \\[3em] 4T_c \dfrac{\cos^2 \left( \pi f T_c \right)}{\left( \pi f T_c \right)^2} \left[ \dfrac{\sin^2 \left( \dfrac{\pi f}{4 f_s} \right)}{\cos \left( \dfrac{\pi f}{2 f_s} \right)} \right]^2 , & k \text{ odd}. \end{array} \right. \tag{2.8}
$$

With $k = \frac{2m}{n} = \frac{2 f_s}{f_c} = \frac{T_c}{T_s}$; again with $T_s$ half the square wave period. The normalized $Rect(1)$ power spectrum is plotted in Fig. 2.2.

The normalized power spectrum for $SinBOC(m, n)$ and $CosBOC(m, n)$ for various $m$ and $n$ are plotted in Fig. 2.3. In comparison to $Rect(n)$ signals that have a dominant main-lobe, $BOC(m, n)$ signals two dominant main-lobes with a separation determined by $m \cdot f_0$. This characteristic enables multiple systems to share the same carrier frequency while reducing co-channel interference due to cross-correlation [7].

A composite normalized power spectrum for $Rect(1)$, $SinBOC(1, 1)$, and $CosBOC(1, 1)$ are plotted in Fig. 2.4 for comparison. Note the increasing signal bandwidths for $Rect(1)$, $SinBOC(1, 1)$, and $CosBOC(1, 1)$, respectively.

Figure 2.5 plots the normalized power profiles, with respect to $f_o$, for $Rect(1)$, $SinBOC(1, 1)$ and $CosBOC(1, 1)$. A receiver must have an input bandwidth of at least $4.5 f_0$, $12 f_0$, and $18 f_0$, respectively, to capture 98% of the total available signal power. Therefore, the disadvantages of

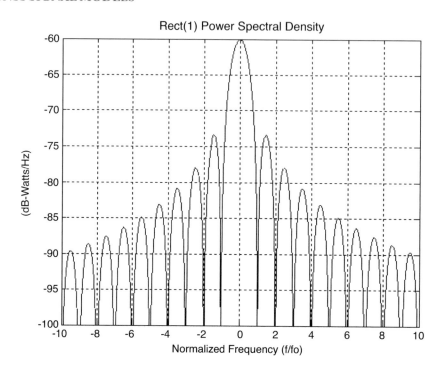

**Figure 2.2:** $S_{Rect(1)}$ normalized power spectrum.

*BOC* signals include an increase in the required receiver bandwidth, sample-rate, and power-consumption.

The *Rect*(1), *SinBOC*(1, 1), and *CosBOC*(1, 1) autocorrelation functions are plotted in Fig. 2.5. The width of the main correlation peak is a measure of time-shift orthogonality (co-herence time) and is inversely proportional to the required signal bandwidth. *CosBOC*(1, 1) has a narrower peak than either *SinBOC*(1, 1) or *Rect*(1). This means that time-shifted replicas of the *BOC*(1, 1) signals decorrelate more than a *Rect* signal given the same time shift, thereby providing an intrinsic level of multi-path mitigation. However, unlike the *Rect* signal with one correlation peak, the *BOC* signals have multiple correlation peaks which can cause false lock points if not properly handled. Both of these points will be explained in detail below.

## 2.2   CHANNEL MODEL AND SIGNAL PROPAGATION

The GNSS signals occupy the same time and bandwidth. The incident signal will be a sum of the signals from the various transmitters present, interference, and additive noise. Let $N$ denote the numbers of transmitters and $L_n$ denote the number of multipath signal components of the $n$-th

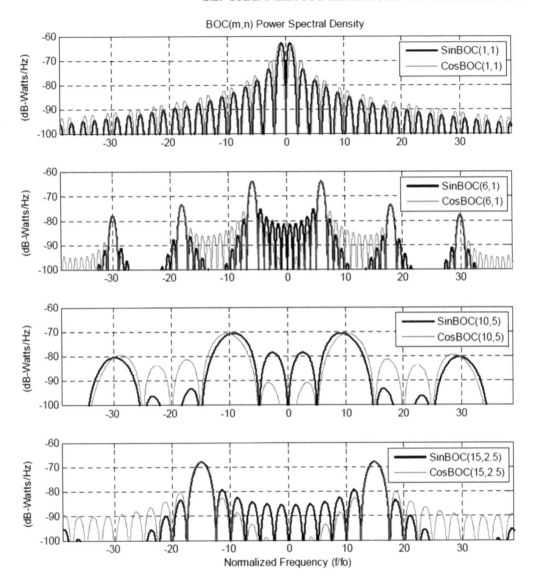

**Figure 2.3:** Normalized power spectrums for various $SinBOC(m, n)$ and $CosBOC(m, n)$ signals.

transmitter. The received signal vector at antenna element, $k$, may be expressed as

$$\mathbf{r}_k(t) = \sum_{n=1}^{N} \sum_{l=1}^{L_n} \mathbf{a}(\theta_{n,l,k}) \alpha_{n,l,k} e^{j\phi_{n,l,k}} x(t - \tau_{n,l,k}) + \mathbf{n}(t), \tag{2.9}$$

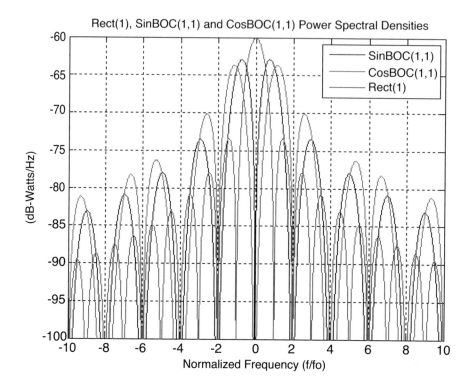

**Figure 2.4:** Composite normalized power spectrum for *Rect*(1), *SinBOC*(1, 1), and *CosBOC*(1, 1).

where $\theta_{n,l,k}$, $\tau_{n,l,k}$, $\alpha_{n,l,k}$, and $\phi_{n,l,k}$ are the angle of arrival, delay time, path attenuation, and path phase of the $l^{\text{th}}$ path of the $n^{\text{th}}$ transmitter, respectively, impinging on antenna element $k$. And $\mathbf{a}(\theta_{n,l})$ is the antenna response vector to a signal impinging upon the element with a direction of arrival (DOA), $\theta_{n,l}$. The receiver noise, $\mathbf{n}(t)$, is Additive White Gaussian Noise (AWGN) with zero mean and variance, $\sigma_n^2$. Figure 2.7 illustrates different path delay scenarios within the channel model. Each ellipsoid represents equal path delay.

When only one transmitter and one receiver antenna element (with an ideal response) are considered, a simplified multipath model for the receiver becomes

$$r(t) = \sum_{l=0}^{L} \alpha_l e^{j\phi_l} x(t - \tau_l) + n(t). \tag{2.10}$$

Let $\alpha_0 \quad 1, \phi_0 \quad 1$, and $\tau_0 \quad 0$ be the normalized line-of-sight signal parameters and parameters with $l > 0$ be with respect to the line-of-sight without loss of generality. From (2.10) it is clear that the ability to resolve multipath or estimate the channel impulse response is dependent upon the signal coherence time (autocorrelation).

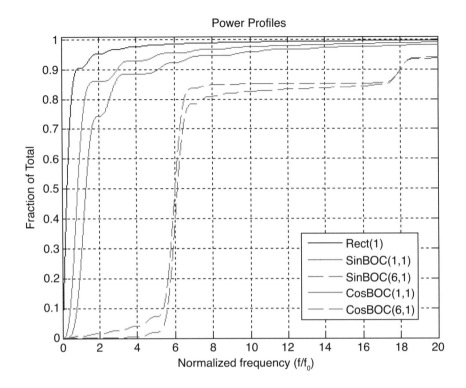

**Figure 2.5:**  Normalized power profiles for *Rect*(1), *SinBOC*(1, 1), and *CosBOC*(1, 1).

All GNSS systems utilize Right Hand Circular Polarized (RHCP) signals since it is more robust to polarization changes that can occur as signals propagate through the Earth's Ionosphere [8]. Propagation models for various conditions are summarized in [9].

## 2.3    RECEIVER STRUCTURES

GNSS receivers utilize a coherent correlator for signal detection and data demodulation. A complex signal model for a DSSS receiver operating on a real signal, $r(t)$, is shown in Fig. 2.8. The analog filter, $H(\omega)$, is located after the last down conversion stage and its bandwidth sets the system performance. It is referred to as the pre-correlation (or pre-detection) filter. The LMQ block includes an automatic gain controller and an adaptive Lloyd-Max Quantizer [8]. To reduce cost, digital sub-sampling is employed so the signal coming out of the analog process block is real with a non-zero digital carrier offset.

The following discussion assumes the signal, $z(t)$, has been down converted to near-baseband using conventional digital carrier tracking methods such as a Costas Loop. Appendix A provides an analysis of the design trade-offs for the digital carrier NCO and phase model suitable

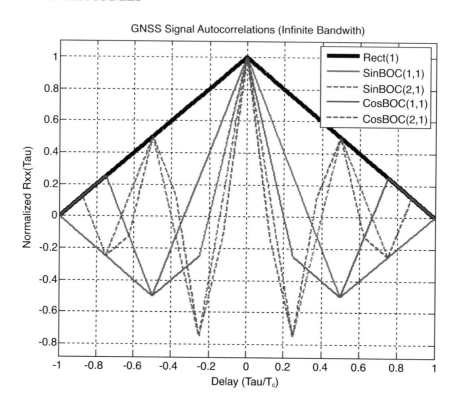

**Figure 2.6:** Normalized autocorrelations.

for GNSS applications and derives the resulting receiver self-noise associated. The remaining discussion concentrates on correlative code phase discriminators used within code tracking control loops.

The code recovery block operates on the complex near-baseband signal, $z(t)$, to track the incoming signal's code phase by driving the code phase error, $\varepsilon$, to zero. The block consists of a phase discriminator, a loop filter, a Numerically Controlled Oscillator (NCO) and a code kernel generator. Two kernels are generated: (1) A model, $\hat{x}(t)$, of the transmitted code sequence, $x(t)$; and (2) a code tracking kernel, $w(t)$. The kernel $\hat{x}(t)$ is correlated with $z(t)$ to form the statistic $R_{Z\hat{X}}(\varepsilon)$. This complex valued statistic is used to form the carrier tracking loop discriminant, $d_\phi(\varepsilon)$ and decode the satellite's data bits. The kernel $w(t)$ is also correlated with $z(t)$ to form the statistic $R_{ZW}(\varepsilon)$. This kernel can be designed with multi-path mitigation properties and used as part of the code phase error discriminant, $d_\varepsilon(\varepsilon)$. The code phase discriminator estimates the phase error between the incoming measured signal and the code model. The filtered phase error is output from the loop filter and provides a delta-phase which is accumulated in the NCO to adjust the

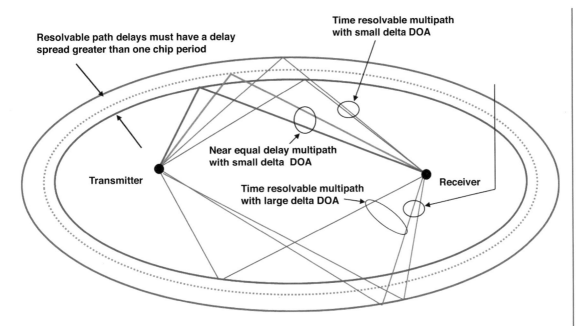

**Figure 2.7:** Propagation path scenarios.

model phase and drive the phase error to zero. The control loop will advance or retard the model in time until it aligns with the incoming signal.

Consider the no-multipath case and using (2.1) to obtain $z(t)$ yields

$$z(t) = \{x(t) + n(t)\} \quad h(t)\, e^{-j\varphi(t)}. \tag{2.11}$$

Solve for the correlation statistic, $R_{\widehat{ZX}}(\varepsilon)$, using a suitable correlation period, $T$, which is determined by the data period and the required tracking dynamics. For GPS L1CA, the correlation period is typically set to the code epoch period of 1 ms. $R_{\widehat{ZX}}(\varepsilon)$ is found by

$$R_{\widehat{ZX}}(\varepsilon) = \frac{1}{T} \int_0^T z(t)\, \hat{x}^*(t - \varepsilon)\, dt, \tag{2.12}$$

and substituting (2.11) into (2.12) and expanding terms yields

$$R_{\widehat{ZX}}(\varepsilon) = \frac{1}{T} \int_0^T \left[ \{x(t) + n(t)\} \quad h(t)\, e^{-j\varphi(t)} \right] \hat{x}^*(t - \varepsilon)\, dt. \tag{2.13}$$

Represent the carrier tracking phase estimate, $e^{-j\varphi(t)}$, as $e^{-j(2\pi f_e t + \varphi_e)}$, then the signaling term for (2.13) becomes

$$R_{\widehat{ZX}}(\varepsilon) = \frac{e^{-j\varphi_e}}{T} \int_0^T \left[ x(t)\, \hat{x}^*(t - \varepsilon)\, e^{-j2\pi f_e t} \right] \quad h(t)\, dt, \tag{2.14}$$

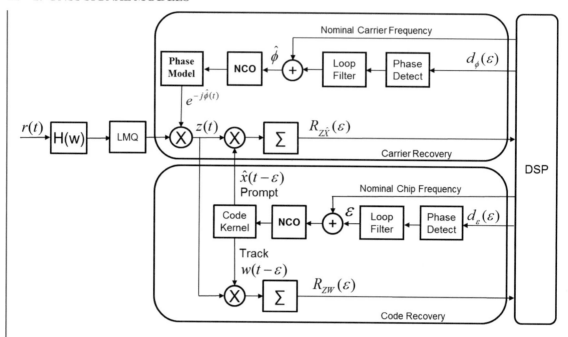

**Figure 2.8:** Receiver complex signal analysis model.

and

$$R_{\widehat{ZX}}(\varepsilon) = \frac{e^{-j\varphi_e}}{T} \frac{e^{-j2\pi f_e t}}{j2\pi f_e t} \int_0^T R_{\widehat{XX}}(\varepsilon) \ h(t),$$ (2.15)

and

$$R_{\widehat{ZX}}(\varepsilon) = \left( R_{\widehat{XX}}(\varepsilon) \ h(t) \right) \sin c\,(f_e T)\, e^{-j\left(2\pi f_e \frac{T}{2} + \varphi_e\right)}.$$ (2.16)

If the receiver is tracking, the frequency and phase error are small, and (2.16) can be simplified to

$$R_{\widehat{ZX}}(\varepsilon) = R_{\widehat{XX}}(\varepsilon) \ h(t).$$ (2.17)

The $R_{\widehat{XX}}(\varepsilon)$ term is the auto-correlation of the spreading sequence. Plots of the auto-correlations and power spectrums for $Rect(n)$, $SinBOC(1,1)$, and $CosBOC(1,1)$ were provided in Fig. 2.3 and Fig. 2.5.

The noise term of (2.13) is

$$N_{\widehat{NX}}(\varepsilon) = \frac{e^{-j\varphi_e}}{T} \int_0^T \left[ n(t)\,\hat{x}^*(t-\varepsilon)\, e^{-j2\pi f_e t} \right] \ h(t) \ dt.$$ (2.18)

Since Gaussian noise is circularly symmetric, the complex rotations of (2.18) have no impact and can be removed. Moreover, since the model code chips are stationary, zero mean, equal probable with constant modulus, the noise term reduces to

$$N_{N\widehat{X}}(\varepsilon) = \frac{1}{T}\int_0^T n(t) \quad h(t) \, dt. \tag{2.19}$$

This yields Gaussian noise with zero mean and variance, $\sigma^2_{N\widehat{X}} = \sigma^2 \, H^{\,2}$.

A similar analysis is conducted to solve for $R_{ZW}(\varepsilon)$ and yields a signaling term of

$$\begin{aligned} R_{ZW}(\varepsilon) &= \frac{1}{T}\int_0^T z(t)\,w^*(t-\varepsilon)\,dt \\ &= (R_{XW}(\varepsilon) \quad h(t)) \sin c\,(f_e T)\, e^{-j\left(2\pi f_e \frac{T}{2}+\varphi_e\right)}. \end{aligned} \tag{2.20}$$

When the receiver is carrier locked, the carrier frequency and phase errors are small, then (2.20) simplifies to

$$R_{ZW}(\varepsilon) = R_{XW}(\varepsilon) \quad h(t). \tag{2.21}$$

The corresponding noise term is

$$N_{NW}(\varepsilon) = \frac{e^{-j\varphi_e}}{T}\int_0^T \left[ n(t)\,W^*(t-\varepsilon)\,e^{-j2\pi f_e t}\right] \quad h(t)\,dt, \tag{2.22}$$

which yields Gaussian noise with zero mean and variance, $\sigma^2_{NW} = \sigma^2 \, WH^{\,2}$. A detailed analysis of the correlation process self-noise for a digital implementation is provided in Appendix A.

The design of the code tracking kernel, $w(t)$, depends on how the statistics $R_{Z\widehat{X}}(\varepsilon)$ and $R_{ZW}(\varepsilon)$ are used to form the code phase discriminator, $d_\varepsilon(\varepsilon)$. The function of the code-phase discriminator is to estimate the signal's time-of-arrival and form a code-phase error estimation. A Maximum Likelihood time-of-arrival estimator is derived below [9, 10]. In general, given an independent identically distributed (iid) observation vector, $x_1, x_2, \ldots, x_N$, which is conditioned on the estimation parameters, $\theta$, the likelihood function is given by

$$L(\theta) = f_\theta(\mathbf{x}) = \prod_{i=1}^N P(x_i \, \theta), \tag{2.23}$$

where $P(\cdot)$ is the conditional pdf. The log-likelihood becomes

$$\Lambda_\theta(\mathbf{x}) = \ln\{L(\theta)\} = \sum_{i=1}^N \ln\{P(x_i \, \theta)\}. \tag{2.24}$$

Therefore, the maximum likelihood is found by

$$\max_\theta \ln\{L(\theta)\}, \tag{2.25}$$

which becomes

$$\frac{\partial}{\partial \theta} \ln \{L(\theta)\} = \sum_{i=1}^{N} \frac{1}{P(x_i \ \theta)} \cdot \frac{\partial}{\partial \theta} P(x_i \ \theta) = 0. \tag{2.26}$$

Now consider time of arrival maximum likelihood estimation with a baseband received signal

$$\mathbf{x}(t) = s(t - \tau) + \mathbf{n}(t), \tag{2.27}$$

where $\mathbf{x}(t)$ is the delayed version of the reference signal, $s(t)$ $\tau$ is the path delay; and $\mathbf{n}(t)$ is additive white Gaussian noise with zero mean and variance, $\sigma^2$. The estimation parameter is $\theta = [\tau]$. The likelihood function is

$$L(\theta) = f_\theta(\mathbf{x}) = \sum_{i=1}^{N} (2\pi\sigma^2)^{-1/2} e^{\frac{-1}{\sigma^2}(x[i] - s[i;\theta])^2}, \tag{2.28}$$

and

$$L(\theta) = f_\theta(\mathbf{x}) = (2\pi\sigma^2)^{-\frac{N}{2}} e^{\frac{-1}{\sigma^2} \sum_{i=1}^{N} (x[i] - s[i;\theta])^2}. \tag{2.29}$$

Note that the reference signal is taken as the distribution's mean. The log-likelihood function is

$$\Lambda_\theta(\mathbf{x}) = -\frac{N}{2} \ln(2\pi\sigma^2) - \frac{1}{\sigma^2} \sum_{i=1}^{N} (x[i] - s[i \ \theta])^2. \tag{2.30}$$

Maximizing $\Lambda_\theta(\mathbf{x})$ with respect to $\theta$ is the same as

$$\widehat{\theta} = \max_\theta \left\{ \frac{1}{\sigma^2} \sum_{i=1}^{N} (x[i] - s[i \ \theta])^2 \right\}, \tag{2.31}$$

and after expanding the summation yields

$$\widehat{\theta} = \max_\theta \left\{ \frac{1}{\sigma^2} \sum_{i=1}^{N} \left( x[i]^2 - 2x[i] s[i \ \theta] + s[i \ \theta]^2 \right) \right\}. \tag{2.32}$$

Note that the reference signal for $Rect(n)$, $SinBOC(m,n)$, and $CosBOC(m,n)$ have constant modulus, so $s[i \ \theta]^2 = 1$, for all $\theta$. This constant modulus assumption does not hold for all $Alt$-$BOC(m,n)$ signals. Given constant modulus,

$$\widehat{\theta} = \max_\theta \left\{ \sum_{i=1}^{N} x[i]s[i \ \theta] \right\}. \tag{2.33}$$

Solving (2.26) using (2.33) for $\widehat{\theta}$ yields

$$\sum_{i=1}^{N} x[i]s[i\ \theta] = \sum_{i=1}^{N} x[i]\frac{\partial}{\partial\theta}s[i\ \theta] = 0, \tag{2.34}$$

with a change of variables, $\frac{\partial}{\partial\theta} = -\frac{\partial}{\partial t}$, then (2.34) becomes

$$\widehat{\theta} = \sum_{i=1}^{N} x[i]\frac{\partial}{\partial t}s[i\ t] = 0. \tag{2.35}$$

Which shows the optimal estimate for the signal delay $\tau$, in the maximum likelihood sense, is found by solving for when the cross correlation of the incoming signal and the derivative of the spreading code reference model is zero. This provides the basis for the delay locked loop (DLL) code phase discriminator with the tracking kernel equal to the code reference derivative [10]

$$\frac{\partial}{\partial t}s[i\ t] \quad w_\delta(t) = \frac{x(t+\delta) - x(t-\delta)}{2\delta}. \tag{2.36}$$

Note as $\delta \quad 0$, then (2.36) better approximates the derivative.

Since the spreading code sequence is known in advance, the DLL is naturally extended to DSSS signals via the $\delta$-delay Early-Late (EML) discriminator described by [12] and [13]. References [13] and [14] analyzed loop performance with respect to $\delta$ for both low and high SNR and it was determined that the $\delta$ values other than $1/2$ could be optimal. Moreover, [15] describes a scheme to vary $\delta$, $(0 < \delta < 1)$, to improve acquisition time and tracking bandwidth.

The code kernel block within Fig. 2.7 generates an EML tracking kernel

$$w_\delta(t) = \frac{\hat{x}(t+\delta) - \hat{x}(t-\delta)}{2}, \tag{2.37}$$

where $\hat{x}(t+\delta)$ and $\hat{x}(t-\delta)$ are the $\delta$-chip early and $\delta$-chip late spreading code model, respectively. Figure 2.9 shows the $\delta = 1/2$ chip early sequence, an on-time (prompt) sequence, a $\delta = 1/2$ chip late sequence, and the resulting correlation kernel, $w_\delta(t)$. Note how the EML signal approximates the derivative of the prompt signal.

The prompt, in this case, represents the signal to track, $z(t)$. Consider a discriminant, $d_\varepsilon(\varepsilon) = R_{ZW}(\varepsilon)$, and no multipath. Then $d_\varepsilon(\varepsilon)$ is found by substituting in (2.37) into (2.20) to yields

$$d_\varepsilon(\varepsilon) = R_{ZW}(\varepsilon) = \frac{1}{T}\int_0^T z(t)\frac{\hat{x}^*(t+\delta-\varepsilon) - \hat{x}^*(t-\delta-\varepsilon)}{2}\,dt, \tag{2.38}$$

and simplifies to

$$d_\varepsilon(\varepsilon) = R_{ZW}(\varepsilon) = \frac{R_{X\widehat{X}}(\varepsilon+\delta) - R_{X\widehat{X}}(\varepsilon-\delta)}{2} \quad h(t). \tag{2.39}$$

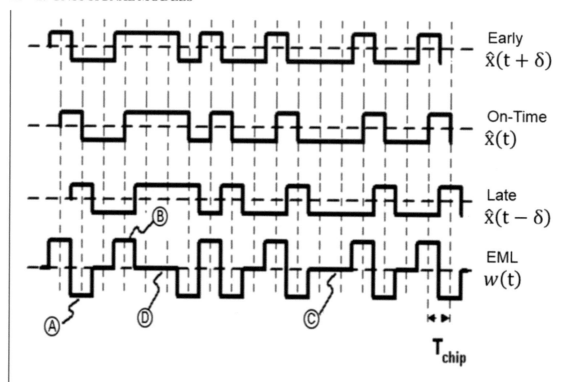

**Figure 2.9:** Generation of the prompt and Early Minus Late (EML) correlation signals for GPS L1CA with $\delta = 1/2\,T_c$.

Note that the code phase discriminant is a linear combination of the spreading code auto-correlation, $R_{\widehat{X}\widehat{X}}(\varepsilon)$. This is expected, since the EML tracking kernel is a linear combination of the spreading code. Figure 2.10 shows the EML correlation tracking kernel, $w(t)$, applicable for $Rect(n)$ signals and the resulting phase-discriminator output, $R_{XW}(\varepsilon)$, for variable $\delta$. Note that $w(t)$ is even symmetric and $R_{XW}(\varepsilon)$ is odd symmetric about $\varepsilon = 0$.

The correlation function is defined such that prompt is the reference and is held stationary. The axis is such that a delay in time alignment is considered positive and to the right. When the signal arrives earlier than the prompt model, $\hat{x}(t)$, its resulting correlation function is shifted to the right since the signal must be delayed before it aligns with prompt. When the signal arrives later than prompt, its correlation function is shifted to the left since it must advance to align with prompt. Note that the EML correlation function crosses zero at $\varepsilon = 0$ when early and late codes are balanced $\delta$-chip on either side of a prompt code. A detailed analysis of the EML discriminator is presented in Appendix A. When the incoming signal arrives early with respect to the model, more energy is present in the early correlation result and a positive error is produced. The positive error advances the reference phase and the EML model moves earlier in time. A negative error

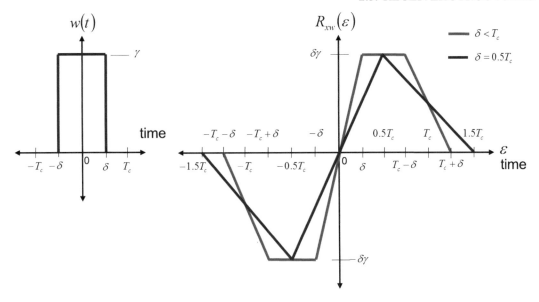

**Figure 2.10:** ML-DLL *Rect(n)* Kernel shape, $w(t)$, and discriminator gain, $R_{XW}(\varepsilon)$.

arises when the incoming signal arrives late and more energy shows up in the late sequence. This delays the reference phase. The error continuously drives the delay-lock-loop, which thereby adjusts the reference phase generator to maintain zero phase error.

Many factors affect the code phase discriminator: Noise, signal amplitude, carrier offset, transition probability, and multipath. All factors will reduce the phase detector gain [9]. The dependence on amplitude can be eliminated by employing an automatic gain controller (AGC). The code tracking loop can function with a small carrier offset which is some fraction of the code rate. However, the carrier offset has the effect of reducing the average signal amplitude which reduces the phase detector gain. The code bit transition probability is set by the code bit sequence. A phase discriminator bias occurs due to multipath.

A normalized ML code tracking kernel for *SinBOC*(1, 1) is shown in Fig. 2.11. Note that the kernel weight at $t = 0.5$ has twice the amplitude as the weights at $t = 0$ and $t = 1.0$ since the *SinBOC*(1, 1) signal has a guaranteed transition at $t = 0.5$ and only a 50% transition probability at $t = 0$ and $t = 1.0$. The resulting code phase discriminant is shown in Fig. 2.12. Note the discriminant is zero at both $\varepsilon = 0$ and $\varepsilon = \pm T_c$ which creates a phase tracking ambiguity. This is characteristic of all *BOC*($m, n$) signals and is due to the additional signal transitions within $g(t)$.

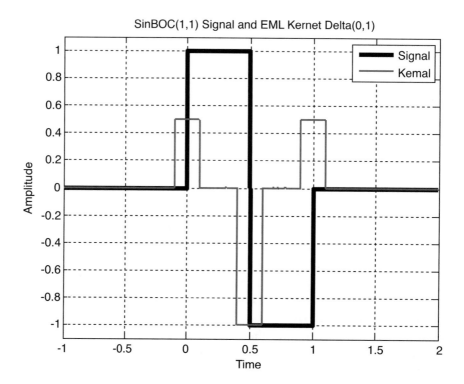

**Figure 2.11:** ML *SinBOC*(1, 1) code tracking kernel, $w(t)$.

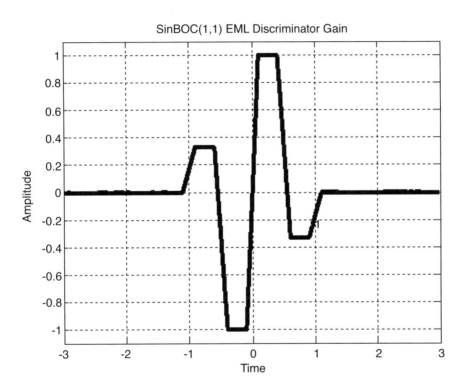

**Figure 2.12:** ML *SinBOC*(1, 1) discriminator gain, $R_{XW}(\varepsilon)$.

# CHAPTER 3

# Existing GNSS Multipath Mitigation Techniques

Chapter 1 showed that multipath is a dominant error source for both stand-alone and differential GNSS receivers. Chapter 2 provided the mathematical models for GNSS signals, multipath channels, and receivers. These models will be used within this chapter to survey existing GNSS multipath mitigation techniques. These techniques can be placed into three categories: (1) Fixed Radiation Pattern Antennas; (2) Code tracking via digital signal processing; and (3) Multiple antenna systems. The concentration will be on signal processing techniques since this provides a cost and power effective solution to both consumer and precision GNSS receivers. Precision GNSS receivers for surveying and small platform autonomous control will also use a high quality fixed radiation pattern antenna. Large platform autonomous control can accommodate the high cost, power, and size of a multiple antenna scheme.

## 3.1 FIXED RADIATION PATTERN ANTENNAS (FRPA)

Recall that all GNSS signals are Right Hand Circular Polarized (RHCP) since it is more robust to polarization changes that can occur as signals propagate through the Earth's Ionosphere [8]. The FRPA design goals are: (1) Maximize the antenna gain for left-handed circular polarized signals in the skyward (up) direction; (2) Minimize the gain for signals impinging upon the antenna from below or with Left Hand Circular Polarization (LHCP) since these are reflected (multipath) signals; and (3) Minimize phase *variation* across elevation and azimuth which manifests as a position dependent variation in satellite range [16]. The FRPA simply mitigates multipath by attenuating any signals impinging upon the antenna from below. Multipath impinging upon the antenna with a positive elevation angle is passed. An FRPA element is pictured in Fig. 3.1 and the corresponding received signal gain pattern vs. elevation angle is shown in Fig. 3.2.

## 3.2 DIGITAL SIGNAL PROCESSING MULTIPATH MITIGATION METHODS

Consider a coherent code phase discriminant defined as

$$d_\varepsilon(\varepsilon) = R_{XW}(\varepsilon). \tag{3.1}$$

Figure 3.1: GNSS fixed radiation pattern antenna element.

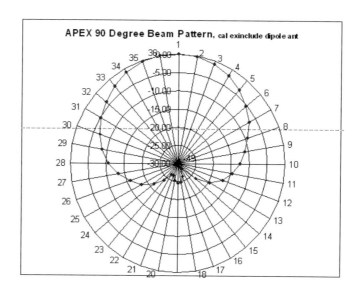

Figure 3.2: GNSS FRPA gain vs. elevation angle (1 = 10°, 36 = 360°, Radius in dB).

Consider the multipath channel represented in (2.10) such that the received signal consists of a direct path and $L-1$ multi-path signals with delay $\tau_i$ and path gain $\alpha_i$ *relative* to the direct path

$$r(t) = \sum_{l=0}^{L} \alpha_l e^{j\phi_l} x(t - \tau_l) + n(t), \tag{3.2}$$

and equivalently,

$$r(t) = x(t) + \sum_{l=1}^{L} \alpha_l e^{j\phi_l} x(t - \tau_l) + n(t). \tag{3.3}$$

Substituting (3.3) into (2.20) and following the computations and assumptions outlined in Chapter 2, the correlation statistics, $\widetilde{R}_{XW}(\varepsilon)$ is

$$d_\varepsilon(\varepsilon) = \widetilde{R}_{XW}(\varepsilon) = R_{XW}(\varepsilon) + \sum_{i=1}^{L} \alpha_i R_{XW}(\varepsilon - \tau_i) + N_{NW}. \tag{3.4}$$

Where the summation term in (3.4) is the phase discriminant distortion due to multipath; the tilde denotes signals that include multipath; and the noise term, $N_{NW}$, is that of (2.22). An EML $\delta = 1/2T_c$ code discriminator output when the received signal, $r(t)$, includes multipath is shown in Fig. 3.3. Note how the multipath creates a discriminant bias such that $d_\varepsilon(\varepsilon) \pm 0$ when $\varepsilon = 0$.

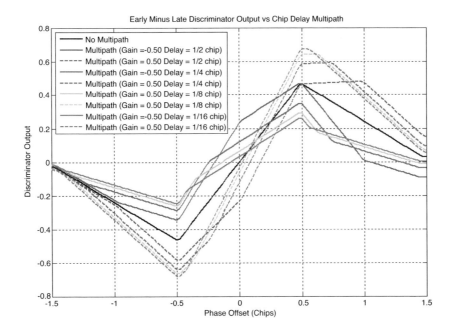

Figure 3.3: EML discriminator multipath performance with $\delta = 1/2T_c$.

Also note that the magnitude of the bias generally increases with multipath delay, $\tau_i$, but depends upon the magnitude and sign of $\alpha_i$. It is important to note that the code phase bias can either be positive or negative. This code tracking timing bias generates a time of arrival estimation bias and consequently a GNSS position bias.

Two primary solution classes exist to remove the multipath distortion via digital signal processing: (1) Parameter estimation of $\alpha_0$, $\phi_0$, and $\tau_0$; and (2) Correlation kernel design of $w(t)$.

Parameter estimators utilize the $R_{Z\widehat{X}}(\varepsilon)$ correlation statistic to perform a deconvolution or a projection onto convex sets (POCS). Substituting (3.3) into (2.12) and following the computations and assumptions outlined in Chapter 2, the correlation statistics, $\widetilde{R}_{Z\widehat{X}}(\varepsilon)$ is

$$\widetilde{R}_{Z\widehat{X}}(\varepsilon) = R_{Z\widehat{X}}(\varepsilon) + \sum_{i=1}^{L} \alpha_i R_{Z\widehat{X}}(\varepsilon - \tau_i) + N_{N\widehat{X}}, \tag{3.5}$$

which can be simplified to

$$\widetilde{R}_{Z\widehat{X}}(\varepsilon) = R_{X\widehat{X}}(\varepsilon) \quad h(t) + \sum_{i=1}^{L} \alpha_i R_{X\widehat{X}}(\varepsilon - \tau_i) \quad h(t) + N_{N\widehat{X}}. \tag{3.6}$$

Where the summation term in (3.6) is the distortion of the autocorrelation due to multipath; the tilde denotes signals that include multipath; and the noise term, $N_{N\widehat{X}}$, is that of (2.18). Representing (3.6) in matrix notation yields

$$\widetilde{\mathbf{R}}_{Z\widehat{X}} = [\mathbf{R}_{X\widehat{X}} \quad h(t)]\mathbf{q} + \mathbf{N}_{N\widehat{X}}, \tag{3.7}$$

where $\mathbf{R}_{X\widehat{X}}$ is an $N$ $N$ sampled autocorrelation matrix that spans $-T_c$ to $T_c$ in intervals of any multiple of the receiver sampling period; $N$ is the number of signal paths to estimate; $\mathbf{q}$ is an $N$ $1$ column vector of estimated path gains, and $\mathbf{N}_{N\widehat{X}}$ is an $N$ $1$ noise vector. The channel impulse response can be estimated as

$$\hat{\mathbf{q}} = \left({}^2\mathbf{I} + \mathbf{R}_{X\widehat{X}}^{T}\mathbf{R}_{X\widehat{X}}\right)^{-1} \mathbf{R}_{X\widehat{X}}^{T}\widetilde{\mathbf{R}}_{Z\widehat{X}}, \tag{3.8}$$

or the solution can be solved by iteration. ${}^2$ is the estimated noise variance. The disadvantage of these solutions are their complexity and scale considering each GNSS receiver processes hundreds of channels and each code phase tracking correlation statistic will be replaced by a $2N + 1$ correlators to provide the sampled autocorrelation function. Moreover, a complex matrix inversion is required every loop update which is about 1 ms for GPS L1CA. A commercial GPS L1CA receiver utilizing this method is the Multipath Estimating Delay Locked Loop (MEDLL) [19].

The Teager-Kaiser (TK) non-linear energy operator has been applied by [27] to the autocorrelation statistic as a means to estimate multipath delay. The real, discrete time TK operator with sample rate, $T_s$, is

$$\Psi_{\delta}[x(n)] = x^2(n) - x(n-1)x(n+1). \tag{3.9}$$

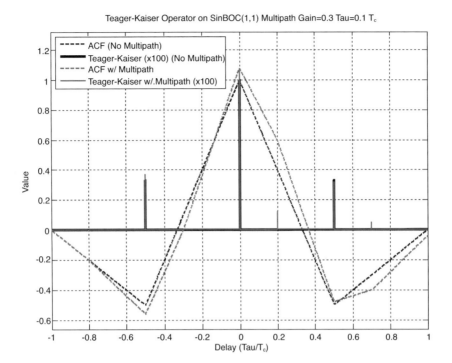

**Figure 3.4:** Teager-Kaiser operator on *SinBOC*(1, 1).

A plot of the TK operator on *SinBOC*(1, 1) with multipath is shown in Fig. 3.4.

Multipath components are resolvable to within the windowed resolution of the sample period. This scheme is more susceptible to noise due to its non-linear operation. Additional multipath mitigation schemes that utilize the sampled autocorrelation or a linear combination of the sampled autocorrelation. The most notable is [20] which included constraints on the discriminant to set minimum gain and resolve the BOC discriminator ambiguity.

The multipath distortion can also be reduced by limiting the ROS for the code tracking kernel, $w(t)$, such that $R_{XW}(\varepsilon - \tau_i) = 0$ for $\tau_i < \delta$ as in [21, 22] and [23]. The Narrow Correlator sets the EML $\delta$ parameter to fewer than 0.5 chips with a limit set by the signal bandwidth (as described in Section 3.3) [13, 14]. The W-Correlator builds a W-shaped correlation waveform for a linear combination of the EML signal [18]. The Strobe Correlator blanks (sets to zero) the correlation waveform when the PN does not transition [19]. The waveform design correlator utilizes a linear combination of EML to produce a variety of correlation waveforms [21]. All of the methods so far build the correlation waveform utilizing a linear combination of EML and are even symmetric. The multipath distortion can be reduced by limiting the ROS for the code tracking kernel $w(t)$ as in the coherent discriminant case. Figure 3.5 shows the EML discriminator multipath performance with $\delta = T_c/16$ for various multipath conditions.

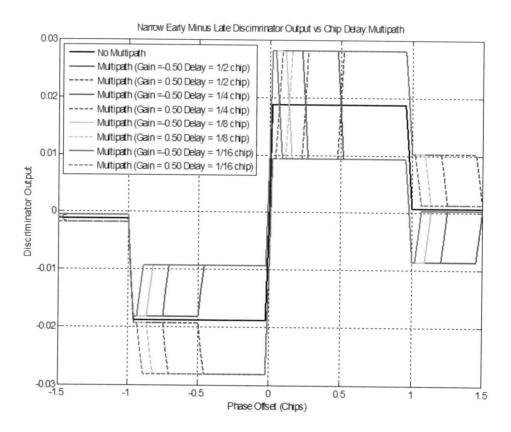

**Figure 3.5:** EML discriminator multipath performance with $\delta = T_c/16$.

These solutions have the disadvantage of narrowing the tracking range and lowering the discriminator gain to $\delta$. Until [5, 6], $w(t)$ has been limited to linear functions of the spreading code which produced an even functioned $w(t)$.

The multipath distortion can also be reduced by designing the code tracking kernel, $w(t)$, to meet specified objectives such as a non-ambiguous *BOC* discriminant [26] or a desired discriminant shape [22]. The unambiguous S-curve shaping technique starts with an S-curve prototype and solves for the linear combination of the spreading code. This technique mitigates medium and long range multipath delay but has poorer results for delay spread of fewer than 0.4 chips. An advantage of this technique is that it generates an unambiguous correlation kernel for *CosBOC*; however, the resulting correlation kernel pulses are narrow and the effects of a band limited pre-correlation filter were not considered. The ROS is greater than 3 chips.

Consider a non-coherent dot-product discriminant defined as

$$d_\varepsilon(\varepsilon) = R_{\widehat{ZX}}(\varepsilon) R_{XW}^*(\varepsilon). \tag{3.10}$$

Substituting (3.10) into the non-coherent discriminant in (3.1) yields

$$
\begin{aligned}
d_\varepsilon(\varepsilon) =& \widetilde{R}_{\widehat{ZX}}(\varepsilon) \widetilde{R}_{XW}^*(\varepsilon) \\
=& \left\{ R_{\widehat{ZX}}(\varepsilon) + \sum_{i=1}^{L_n} \alpha_i\, R_{\widehat{ZX}}(\varepsilon - \tau_i) \right\} \\
& \cdot \left\{ R_{XW}^*(\varepsilon) + \sum_{i=1}^{L_n} \alpha_i^*\, R_{XW}^*(\varepsilon - \tau_i) \right\},
\end{aligned}
\tag{3.11}
$$

and expanding yields

$$
\begin{aligned}
d_\varepsilon(\varepsilon) =& R_{\widehat{ZX}}(\varepsilon) R_{XW}^*(\varepsilon) + R_{\widehat{ZX}}(\varepsilon) \sum_{i=1}^{L_n} \alpha_i^*\, R_{XW}^*(\varepsilon - \tau_i) \\
& + R_{XW}^*(\varepsilon) \sum_{i=1}^{L_n} \alpha_i\, R_{\widehat{ZX}}(\varepsilon - \tau_i) \\
& \pm \sum_{i=1}^{L_n} \alpha_i{}^2 R_{\widehat{ZX}}(\varepsilon - \tau_i) R_{XW}^*(\varepsilon - \tau_i).
\end{aligned}
\tag{3.12}
$$

Similar to the coherent discriminant, the multipath distortion can be reduced by limiting the ROS for the code tracking kernel, $w(t)$, such that $R_{XW}(\varepsilon - \tau_i) = 0$ for $\tau_i < \delta$.

CHAPTER 4

# Summary

Global navigation satellite systems (GNSS) estimate position based upon time-of-arrival estimates obtained by tracking a satellite's code and carrier phase. Multipath is a dominant error source since it corrupts the signal phase estimates with a time-varying bias. This booklet described the theory and design of a GNSS receiver. More specifically in the booklet we provided the principles of GPS systems and we gave a brief literature review of existing GNSS time of arrival estimation methods. In addition, we described the basics of a GNSS receiver model. GNSS signal models were presented and multipath mitigation techniques were described for various multipath conditions. Appendices are included in the booklet with derivations of some of the basics on early minus late code synchronization methods. Appendices also include details on the numerically controlled oscillator and its properties. MATLAB™ code is provided in a file with instructions in Appendix C to enable readers to run simple GNSS receiver simulations.

# APPENDIX A

# Early Minus Late Discriminator Derivation

An Early Minus Late (EML) code synchronization algorithm is developed and its performance with respect to data transition density, carrier frequency offsets, and symbol timing frequency offsets is analyzed. Furthermore, a symbol lock detector is proposed and its performance investigated. The symbol synchronization algorithm operates on complex, near base-band data which consists of multiple samples per symbol to determine the symbol timing phase. Figure A.1 illustrates the early-late code phase tracking timing diagram which is used to calculate the early-late code discriminator.

Where $A$ is the code chip amplitude; $T$ is the true code chip period; $\widehat{T}$ is the current estimate of the code chip period and will deviate from truth due to doppler and self-noise; $\Delta$ is the timing error between the start of the next true code chip and the start of the next estimated code chip; and finally, $\delta$ is the early and late code spacing expressed as a fraction of the estimated code chip period. Consider a code sequence with transition probability, $p$; then the probability that the next code chip has the same amplitude as the current code chip (non-transition probability) is $(1 - p)$. A correlation metric is then calculated for the prompt, early, and late channels considering both the chip transition and chip non-transition case.

When no transition occurs, the expected value of the prompt correlator is $A$. This occurs with probability, $(1 - p)$. When a transition occurs, the expected absolute value of the prompt correlator is given by,

$$E\left[\text{prompt}\right] = \frac{-1}{\widehat{T}} \int_{T-\Delta}^{T} A dt + \frac{1}{\widehat{T}} \int_{T}^{T-\Delta+\widehat{T}} A dt, \tag{A.1}$$

$$E\left[\text{prompt}\right] = \frac{-A\left[T - (T - \Delta))\right]}{\widehat{T}} + \frac{A\left[T - \Delta + \widehat{T}\right]}{\widehat{T}}, \tag{A.2}$$

$$E\left[\text{prompt}\right] = \frac{-A\Delta}{\widehat{T}} + \frac{A\left[\widehat{T} - \Delta\right]}{\widehat{T}}, \tag{A.3}$$

$$E\left[\text{prompt}\right] = \frac{A\left[\widehat{T} - 2\Delta\right]}{\widehat{T}}. \tag{A.4}$$

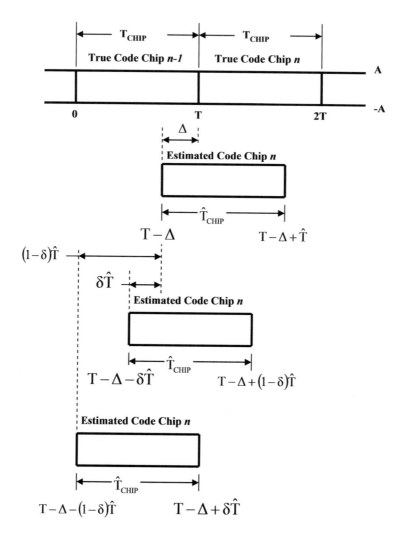

**Figure A.1:** Early-late code phase tracking timing diagram.

Finally, combining the transition and non-transition expectations yields

$$E\left[\text{prompt}\right] = A(1-p) + \frac{A\left[\widehat{T} - 2\Delta\right]p}{\widehat{T}}, \tag{A.5}$$

$$E\left[\text{prompt}\right] = A - Ap + Ap - \frac{2A\Delta p}{\widehat{T}}, \tag{A.6}$$

$$E\left[\text{prompt}\right] = A - \frac{2A\Delta p}{\widehat{T}}. \tag{A.7}$$

The expected absolute value of the non-transition early channel is simply $A$ and occurs with probability $(1-p)$. The expected absolute value of the early channel for the transition case is

$$E\left[\text{early}\right] = \frac{-1}{\widehat{T}}\int_{T-\Delta-\delta\widehat{T}}^{T} A\,dt + \frac{1}{\widehat{T}}\int_{T}^{T-\Delta+(1-\delta)\widehat{T}} A\,dt, \tag{A.8}$$

$$E\left[\text{early}\right] = \frac{-A\left[T - \left(T - \Delta - \delta\widehat{T}\right)\right]}{\widehat{T}} + \frac{A\left[T - \Delta + (1-\delta)\widehat{T} - T\right]}{\widehat{T}}, \tag{A.9}$$

$$E\left[\text{early}\right] = \frac{-A\left(\Delta + \delta\widehat{T}\right)}{\widehat{T}} + \frac{A\left[(1-\delta)\widehat{T} - \Delta\right]}{\widehat{T}}, \tag{A.10}$$

$$E\left[\text{early}\right] = \frac{A\widehat{T}\,(1-2\delta) - 2A\Delta}{\widehat{T}}. \tag{A.11}$$

The expected absolute value of the non-transition late channel is simply $A$ and occurs with probability $(1-p)$. The expected absolute value of the late channel for the transition case is

$$E\left[\text{late}\right] = \frac{-1}{\widehat{T}}\int_{T-\Delta-(1-\delta)\widehat{T}}^{T} A\,dt + \frac{1}{\widehat{T}}\int_{T}^{T-\Delta+\delta\widehat{T}} A\,dt, \tag{A.12}$$

$$E\left[\text{late}\right] = \frac{-A\left[T - \left(T - \Delta - (1-\delta)\widehat{T}\right)\right]}{\widehat{T}} + \frac{A\left[T - \Delta + \delta\widehat{T} - T\right]}{\widehat{T}}, \tag{A.13}$$

$$E\left[\text{late}\right] = \frac{-A\left[\Delta + (1-\delta)\widehat{T}\right]}{\widehat{T}} + \frac{A\left(\delta\widehat{T} - \Delta\right)}{\widehat{T}}, \tag{A.14}$$

$$E\left[\text{late}\right] = \frac{A\widehat{T}\,(1-2\delta) - 2A\Delta}{\widehat{T}}. \tag{A.15}$$

The discriminator metric is

$$J_\Delta = E\left[\text{early}\right] - E\left[\text{late}\right]. \tag{A.16}$$

When no transition occurs, the metric is

$$J_\Delta = A(1-p) - A(1-p) = 0, \tag{A.17}$$

which makes intuitive sense since no timing information is contained in a non-transitioning sequence. When transitions occur, (A.11), (A.15), and (A.16) are combined to form the metric

$$J_\Delta = \frac{p\left[A\widehat{T}(1-2\delta)-2A\Delta\right]}{\widehat{T}} - \frac{p\left[A\widehat{T}(1-2\delta)-2A\Delta\right]}{\widehat{T}}, \tag{A.18}$$

$$J_\Delta = \frac{p\left[\left(A\widehat{T}(1-2\delta)-2A\Delta\right)-\left(A\widehat{T}(1-2\delta)-2A\Delta\right)\right]}{\widehat{T}}, \tag{A.19}$$

$$J_\Delta = \frac{p\left[\left(A\widehat{T}(1-2\delta)-A\widehat{T}(2\delta-1)-2A\Delta+2A\Delta\right)\right]}{\widehat{T}}, \tag{A.20}$$

$$J_\Delta = \frac{4Ap\delta\widehat{T}}{\widehat{T}}. \tag{A.21}$$

Many factors affect the S-curve: Noise, carrier offset, signal amplitude, and transition probability. All factors will reduce the phase detector gain, which directly reduces the open loop gain. The dependence on amplitude and transition density can be eliminated by employing an automatic gain controller (AGC) and scrambler respectively.

The symbol tracking loop can function with a small carrier offset which is some fraction of the symbol rate. However, the carrier offset has the effect of reducing the average signal amplitude which reduces the phase detector gain.

# APPENDIX B

# Carrier NCO Analysis and Correlation Self Noise

Coherent spread spectrum receives required knowledge of all signal phases to perform optimal detection. The required phase information includes the carrier phase and frequency as well as the spreading code sequence, phase, and frequency. Knowledge of signal phase is obtained by comparing the incoming signal to a precisely controlled replica; and then adjusting the replica to minimize the error between it and the true signal. A numerically controlled oscillator (NCO) is used to generate and precisely control the replica. The GNSS carrier NCO is used within the carrier tracking loop to translate the signal to baseband, track out up to $\pm 5$ KHz of doppler frequency shift, and precisely measure the carrier cycles and phase. A detailed general analysis of NCO properties is found in Analog Devices tech document. The following discussion provides specific values that are found in some commercial GPS L1CA receivers.

Consider the $\cos(\theta)$ function. The phase at time, $t$, is given by $\theta = \theta_o + 2\pi f t$ radians. Where $\theta_o$ is the starting phase and $f$ is the frequency. Now consider a digital representation with a sample period, $T_{sample}$, and a constant frequency over the sample period, denoted $\theta$,

$$\theta = \theta_o + 2\pi\theta \cdot N T_{sample}, \tag{B.1}$$

with $t = N \cdot T_{sample}$. A 1$^{st}$ order difference equation is developed from

$$\theta = \theta_o + 2\pi\theta \cdot (N-1) T_{sample} + 2\pi\theta \cdot T_{sample},$$
$$\theta = \theta_o + \left(2\pi\theta T_{sample}\right) \cdot (N-1) + \left(2\pi\theta \cdot T_{sample}\right), \tag{B.2}$$
$$\theta = \theta_o + \sum_0^N \left(2\pi\theta T_{sample}\right).$$

Let $M = 2\pi\theta \cdot T_{sample}$ which defines a phase increment that is accumulated every sample period. This notion of a phase accumulator forms the basis for the NCO. The NCO structure is illustrated in Fig. B.1 and consists of two components: (1) A phase accumulator and (2) a phase to amplitude converter.

The phase increment register $M$ holds the phase accumulation that will occur every $T_{sample}$ seconds. The accumulated phase is stored in a finite length accumulator ($A$ bits) which provides

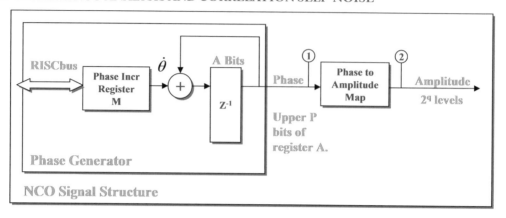

**Figure B.1:** NCO signal structure and local oscillator generation.

a phase resolution of $2\pi/2^A$ radians. The output frequency of the NCO is given by

$$f_o = \frac{f_s}{2^A} \cdot M, \tag{B.3}$$

where $f_s = 1/T_{sample}$. The smallest output frequency adjustment is $f_o = f_s/2^A$ which corresponds to $M = 1$.

The phase is then converted to the appropriate amplitude utilizing a phase to amplitude look up table or a CORDIC processor. Only a phase to amplitude look up table will be considered. A look up table scheme naturally quantizes both the *phase* (address) and the *amplitude* (data): Only the upper $P$ bits of the phase accumulator, $A$, are used to address the $2^P$ amplitude entries, where each entry is represented by $Q$ bits. The quantization and truncation of the phase accumulator, phase address, and amplitude data is a non-linear and noisy process. Consequently, the power spectrum of $f_o$ is not a pure tone, but may include considerable phase noise.

Consider the following NCO common to GPS receivers: $A = 27$ bits, $P = 3$ bits, $B = 2$ bits, $M = $ 0x01F7B1B9, and $f_s = 40.0/7 = 5.714285$ MHz.

The phase accumulator value over time is plotted in Fig. B.2 for a constant $M$. A constant $M$ produces a phase ramp with the slope equal to the NCO output frequency.

Since the accumulator has finite length and will overflow, the accumulator phase is calculated modulo $2^A$. Consequently, the phase may not start at identically zero every cycle, but the zero-phase state will be periodic. The periodicity of the accumulator values creates a spectral tone at $fs/GRR$ where $GRR$ is the grand repetition rate. The $GRR$ is given by

$$GRR = 2^A/GCD(M_b, 2^A), \tag{B.4}$$

where $M_b$ is the $M$ register bit length ($M_b = \log 2(M)$), $GCD$ is the Greatest Common Denominator of $M_b$ and $2^A$. The spectral tone, with the power given by (B.5) should either be positioned

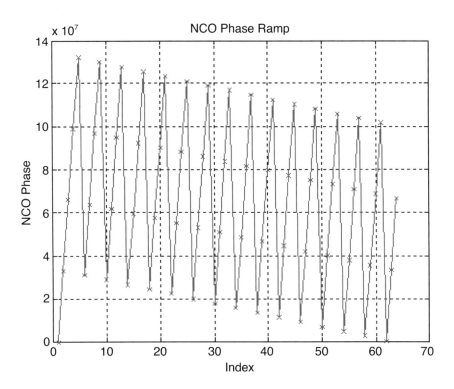

**Figure B.2:** NCO phase accumulator values.

outside the SOI band or at a sufficiently low level so it does not degrade system performance. For this example the *GRR* is 62 samples.

The upper $P$ bits of the phase word are utilized to generate the $2^P$ addresses into the amplitude table. This phase truncation can generate periodic table addresses as illustrated in Fig. B.3, which plots the $8 = 2^{P=3}$ table addresses as a function of the sample index (time). The periodicity of the phase address for the example is one-half the grand repetition rate and is clearly visible within Fig. B.3. Consequently, spectral lines are visible in the signal of interest power spectrum with a power level

$$PTSM = -6.02P \text{ for } A - P > 4 \text{ dBc.} \tag{B.5}$$

For the example with $A = 27$ and $P = 3$, phase truncation will result in spurs of no more than $-18$ dBc regardless of the tuning word.

Moreover, the period will have the effect of modulating the NCO amplitude. The NCO phase to amplitude map is plotted in Fig. B.4 for both the sine and cosine function. The amplitude is quantized to 2 bits for 4 possible levels. The phase is mapped such that 30% of the phases are mapped to the maximum amplitude ($+/-2$ for the example) and the remaining 70% are

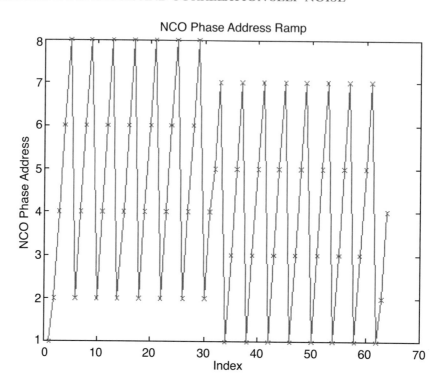

**Figure B.3:** Periodicity of the phase address due to phase truncation to $P$ bits.

$+/-1$. The time domain cosine sequence using the example NCO parameters is illustrated in Fig. B.5. Note the discontinuity at sample index 31. Prior to sample index 31 the amplitude exhibits "concave" distortion toward the left vertical axis. After sample index 31 the distortion converts to a "concave" distortion toward the right vertical axis.

The NCO power spectral density is plotted in Fig. B.6 for the cosine amplitude map. The NCO has been tuned to 1.405 MHz by setting the phase increment, $M$, to $33,010,105$ (0x01F7B19). Note that the largest spur is approximately $-18$ dBc as predicted by (B.5). Selection of the $A$, $M$, $P$, and $Q$ parameters are critical to obtain the desired NCO output frequency and tuning resolution while minimizing the required hardware (bits) and phase noise. It is desirable to select $A$ and $M$ to obtain the highest *GRR* to minimize spurs. The complex PSD is plotted in Fig. B.7 for the complex (cosine and sine) NCO output.

The carrier NCO contributes to the correlation self-noise which sets the 0 dB SNR noise floor and determines the carrier and code lock acquisition and tracking thresholds. With perfect

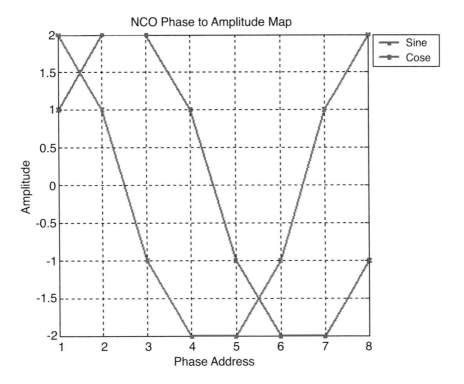

**Figure B.4:** Phase to amplitude map: $P = 3$, $Q = 2$.

code alignment, the correlation process has the form

$$z = \sum_{i=0}^{N_{EPOCH}} x_i a_i, \qquad (\text{B.6})$$

where $x_i$ is the input sample and $a_i$ is the carrier NCO amplitude. The self-noise is a function of the sample and NCO probability distribution functions. The correlation process has zero mean self-noise, $\mu_z = 0$, which is guaranteed by the input AGC and the NCO design. The self-noise power, for a single correlator channel ($I$ or $Q$), over one epoch is

$$\sigma^2 = E\left[z \cdot z^*\right] - \underbrace{\mu_z}_{\equiv 0}. \qquad (\text{B.7})$$

Substituting (B.6) into (B.7) yields

$$\sigma^2 = E\left[\left(\sum_{i=0}^{N_{EPOCH}} x_i a_i\right)\left(\sum_{j=0}^{N_{EPOCH}} x_j a_j\right)\right]. \qquad (\text{B.8})$$

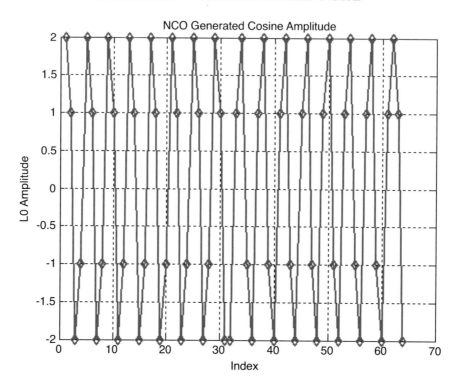

**Figure B.5:** NCO generated cosine amplitude: $A = 27$, $P = 3$, $Q = 2$.

Since the samples are modulated and therefore periodic, the expectations vanish when $i \neq j$. The expectation when $i = j$ is given by

$$\sigma^2 = E\left[\left(\sum_{i=0}^{N_{EPOCH}} x_i^2 a_i^2\right)\right]. \tag{B.9}$$

Let $P_x$ be the Input sample quantization level probability density function. It is set by the number of ADC bits, $Q$, and the automatic gain controller (AGC). The input samples are quantized to $N_x = 4$ levels, $\{-3, -1, +1, +3\}$. The AGC tracking profile sets the sample distribution of 70% for $\pm 1$ samples and 30% for $\pm 3$ samples based upon a 2 bit Lloyd-Max quantizer.

Let $P_a$ be the Carrier NCO quantization level probability density function. The carrier NCO utilizes the upper three bits, $P = 3$, to map eight phases, 23, into the associated amplitude. The amplitudes are quantized to 4 different levels. The phase to amplitude map is given by $\{2, 1, -1, -2, -2, -1, 1, 2\}$. The $M = $ 0x01F7B1B9 tuning generates a complete carrier cycle in fewer than $N_a = 8$ updates, therefore, not every phase amplitude is used each carrier cycle. However, the tuning word and NCO phase accumulator were designed such that every phase has

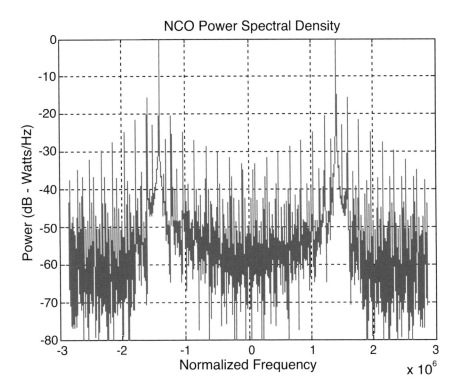

**Figure B.6:** NCO power spectral density: $A = 27$, $P = 3$, $Q = 2$, $M = 0x01F7B1B9$.

equal probability of use over one code epoch. The expectation is removed by

$$\sigma^2 = \sum_{m=1}^{N_x} P_x \sum_{k=1}^{N_a} P_a \left( \sum_{i=0}^{N_{EPOCH}} x_i^2 a_i^2 \right), \tag{B.10}$$

and

$$\sigma^2 = \begin{aligned} & 0.35 \sum_{k=1}^{8} \tfrac{1}{8} \left( \sum_{i=0}^{N_{EPOCH}} 1^2 a_i^2 \right) + 0.35 \sum_{k=1}^{8} \tfrac{1}{8} \left( \sum_{i=0}^{N_{EPOCH}} (-1)^2 a_i^2 \right) + \\ & 0.15 \sum_{k=1}^{8} \tfrac{1}{8} \left( \sum_{i=0}^{N_{EPOCH}} 3^2 a_i^2 \right) + 0.15 \sum_{k=1}^{8} \tfrac{1}{8} \left( \sum_{i=0}^{N_{EPOCH}} (-3)^2 a_i^2 \right) \end{aligned}. \tag{B.11}$$

Simplifying yields

$$\sigma_z^2 = (2 \cdot (0.35 \cdot 1.25) + 2 \cdot (0.15 \cdot 11.25)) \cdot N_{EPOCH} = 48,571. \tag{B.12}$$

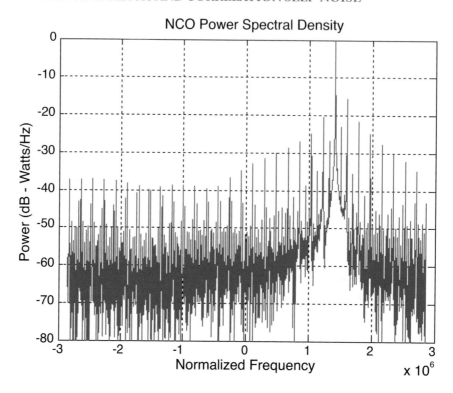

**Figure B.7:** NCO complex power spectral density: $A = 27$, $P = 3$, $Q = 2$, $M = 0x01F7B1B9$.

When considering both $I$ and $Q$ arms, the noise power is

$$\left(I^2 + Q^2\right) = 2\sigma_z^2 = 97,142. \tag{B.13}$$

APPENDIX C

# Sample MATLAB™ Code for Simulating Multipath Effects

## INSTRUCTIONS

To regenerate Fig. 3.3 and Fig. 3.5, please run `Chapter3.m` file. The m file contains the following functions:

1. `gps_signal_gen.m`: generates GPS signal;

2. `SetPhaseBreakPoints.m`: set phase boundaries to describe all the discriminants;

3. `DiscrimSCurve.m`: generate discriminator S-curves given input signal vector which contains either filtered GPS signal and/or signal with multipath; and

4. `ZeroCrossing.m`: find two points of line that crosses $y = 0$.

# References

[1] C. Counselman, "Multipath-Rejecting GPS Antennas," *Proc. of the IEEE*, vol. 87, no. 1, January 1999. DOI: 10.1109/5.736343. 1

[2] L. I. Basilio, J. T. Williams, D. R. Jackson, and M. A. Khayat, "A Comparision Study of a New GPS Reduced-Surface-Wave Antenna," *Antennas Wireless Propag. Lett.*, vol. 4, pp. 233–236, 2005. DOI: 10.1109/LAWP.2005.851105. 1, 4

[3] J. J. Jr. Spilker and D. T. Magill, "The Delay-Lock Discriminator-An Optimum Tracking Device," *Proc. IRE*, vol. 49, no. 9, pp. 1403–1416, 1961. DOI: 10.1109/JRPROC.1961.287899. 6, 7

[4] J. J. Jr. Spilker, "Delay-Lock Tracking of binary signals," *IEEE Trans. Space Electron. Telemetry*, vol. SET-9, p. 18, March 1963. DOI: 10.1109/TSET.1963.4337590. 11, 12

[5] W. J. Gill, "A comparison of binary delay-lock loop implementations," *IEEE Trans. Aerospace Electron. Syst.*, vol. AES-2, pp. 415–424, July 1966. DOI: 10.1109/TAES.1966.4501791. 11, 36

[6] A. Polydoros and C. Weber, "Analysis and Optimization of Correlative Code-Tracking Loops in Spread-Spectrum Systems," *IEEE Trans. on Comm.*, vol. 33, no. 9, pp. 30–43, January 1985. DOI: 10.1109/TCOM.1985.1096193. 11, 12, 36

[7] W. J. Hurd and T. O. Anderson, "Digital Transition Tracking Symbol Synchronizer for LOW SNR Coded Systems," *IEEE Trans. on Comm. Tech.*, vol. COM-18, no. 2, April 1970. DOI: 10.1109/TCOM.1970.1090332. 13, 14, 15

[8] J. K. Holmes and C. C. Chen, "Acquisition Time Performance of PN Spread-Spectrum Systems," *IEEE Trans. on Comm.*, vol. COM-25, no. 8, pp. 778–784, August 1977. DOI: 10.1109/TCOM.1977.1093913. 19, 31

[9] Simon, *IEEE Trans. on Comm. Tech.*, 1970. 19, 23, 27

[10] J. K. Holmes, *Coherent Spread Spectrum Systems*. Malabar, FL: Krieger Publishing Company, 1990. 23, 25

[11] S. Miller, "Sample Clock to Chip Rate Ratio and its Effect on Code Estimation Bias," Unpublished, 2004.

[12] E. D. Kaplan (editor), *Understanding GPS: Principles and Applications.* Artech House Telecommunication Library, 1996. 25

[13] B. Townsend and P. Fenton, "A Practical Approach to the Reduction of Pseudorange Multipath Errors in a L1 GPS Receiver," in *Proc. of ION GPS-94*, Salt Lake City, 1994, pp. 143–148. 25, 35

[14] L. Garin and J. Rousseau, "Enhanced Strobe Correlator Multipath Rejection for Code & Carrier," in *Proc. of ION GPS-97*, Kansas City, 1997, pp. 559–568. 25, 35

[15] V. A. Veitsel, A.V. Zhdanov, and M.I. Zhodzishsky, "The Mitigation of Multipath Errors by Strobe Correlators in GPS/GLONASS Receivers," *GPS Solut.*, vol. 2, no. 2, pp. 38–45, 1998. DOI: 10.1007/PL00000035. 25

[16] B. Schnaufer and G. McGraw, "A Peak Tracking/Measurement Compensation Multipath Mitigation Technique," in *Proc. of the IEEE Position, Location and Navigation Symposium (PLANS 2000)*, San Diego, 2000. 31

[17] L. Weill, "A GPS Multipath Mitigation by Means of Correlator Reference Waveform Design," in *Proc. Institute of Navigation National Technical Meeting*, 1997, pp. 197–206.

[18] J. K. Ray, "Mitigation of GPS Code and Carrier Phase Multipath Effects using a Multi-Antenna System," University of Calgary, Calgary, Ph.D. Thesis UCGE Report No. 20136, 2000. 35

[19] M. S. Braasch, "Isolation of GPS Multipath and Receiver Tracking Errors," *Navigation: Journal of the Institute of Navigation*, vol. 41, no. 4, Winter 1994–1995. 34, 35

[20] M. L. Whitehead and S. R. Miller, "Unbiased Code Phase Discriminator," US 6,744,404 B1, June 1st, 2004. 35

[21] J. Max, "Quantizing for Minimum Distortion," *IRE Trans. on Information Theory*, pp. 7–12, March 1960. DOI: 10.1109/TIT.1960.1057548. 35

[22] M. L. Whitehead and S. R. Miller, "Unbiased Code Phase Discriminator," US 8,000,381 B2, August 15, 2011. 35, 36

[23] R. D. Van Nee, "Spread Spectrum Code and Carrier Synchronization Errors Caused by Multipath and Interference," *IEEE Trans. on Aerospace and Electronic Systems*, vol. 29, no. 4, pp. 1359–1365, October 1993. DOI: 10.1109/7.259541. 35

[24] A. J. Van Dierendonck, P. Fenton, and T. Ford, "Theory and Performance of Narrow Correlator Technology in GPS Receiver," *Navigation, J. Inst. Navi.*, vol. 39, no. 3, pp. 265–283, 1992. DOI: 10.1002/j.2161-4296.1992.tb02276.x.

[25] B. Townsend, P. Fenton, K. Van Dierendonck, and R. D. Van Nee, "L1 Carrier Phase Multipath Error Reduction Using MEDLL Technology," in *Proc. of ION GPS-95*, Palm Spring, 1995, pp. 1539–1544.

[26] F. M. Sousa, F. D. Nunes, and J. M. Leitao, "Strobe Pulse Design for Multipath Mitigation in BOC GNSS Receivers," in *Position, Location, And Navigation Symposium, 2006 IEEE/ION*, 2006, pp. 348–355. DOI: 10.1109/PLANS.2006.1650622. 36

[27] X. Chen, F. Dovis, S. Peng, and Y. Morton, "Comparative Studies of GPS Multipath Mitigation Methods Performance," *IEEE Trans. Aerospace Electron. Syst.*, vol. 49, no. 3, pp. 1555–1568, July 2013. DOI: 10.1109/TAES.2013.6558004. 9, 34

[28] D. Margaria, E. Falletti, A. Bagnasco, and F. Parizzi, "A New Open-loop Multipath Mitigation Strategy Suitable to Modern GNSS Signals," *IEEE/ION Position, Location, and Navigation Symposium*, May 2014. DOI: 10.1109/PLANS.2014.6851418.

[29] H. Viet, C. Ngoc and K. Van, "A Nonlinear Method of Multipath Mitigation for New GNSS Signals," *2014 IEEE Fifth Inter. Conf. Comm. and Electr.*, August 2014. DOI: 10.1109/CCE.2014.6916687.

[30] B. Schipper, "Multipath Detection and Mitigation Leveraging the Growing GNSS Constellation," *IEEE/ION Position, Location, and Navigation Symposium*, May 2014. DOI: 10.1109/PLANS.2014.6851449.

[31] Y. Kamatham, B. Kinnara, and M. Kartan, "Mitigation of GPS Multipath Using Affine Combination of Two LMS Adaptive Filters," *IEEE Inter. Conf. Signal Proc., Informatics, Comm. and Energy Systems*, February 2015. DOI: 10.1109/SPICES.2015.7091375. 9

[32] S. Miller, X. Zhang, and A. Spanias, "A New Asymmetric Correlation Kernel for GNSS Multipath Mitigation," *Sensor Signal Proc. for Defense Conf., IEEE*, September 2015. DOI: 10.1109/SSPD.2015.7288498. 9

[33] A. Spanias, *Digital Signal Processing: An Interactive Approach*. $2^{nd}$ ed., ISBN 978-1-4675-9892-7, Morrisville, NC: Lulu Press On-demand Publishers, May 2014. 9

[34] J. Foutz, A. Spanias, and M. Banavar, "Narrowband Direction of Arrival Estimation for Antenna Arrays," Synthesis Lectures on Antenna, San Rafael, CA: Morgan & Claypool Publishers, August 2008. DOI: 10.2200/S00118ED1V01Y200805ANT008.

[35] S. Miller, "Multipath Mitigating Correlation Kernels," Ph.D. Thesis, Arizona State University, December 2013. 9

# Authors' Biographies

## STEVEN MILLER

**Steven Miller** obtained his Bachelors degree of the University of Wisconsin and the Masters and Ph.D. degrees from Arizona State University. He is the founder and CEO of Aperio DSP. He works on the design of algorithms and software tools for GPS, MSK, PSK, AM and FM receivers. He served as a VP Engineering and Director of Technology at Hemisphere GPS. He previously worked at Sicom, Honeywell and Fairchild as a member of the engineering staff. His research interests are in DSP, Digital Communications, GPS systems and their implementations on various platforms including FPGAs. He co-developed advanced multi-frequency GNSS receiver including the analog frequency plan, digital baseband architecture, signal processing chip with hundreds of channels, and software based signal tracking algorithms. He has published in IEEE conferences and journals and he is the co-author of several patents.

## XUE ZHANG

**Xue Zhang** joined the School of Electrical, Computer, and Energy Engineering at Arizona State University for her Ph.D. degree in Fall 2010, and she is currently a Ph.D. candidate under the supervision of Prof. Andreas Spanias and Prof. Cihan Tepedelenlioglu. Her research interests are in the areas of detection and estimation theory, localization in wireless sensor networks, and optimization. Her current research is localization in wireless sensor networks. She has served as a reviewer for IEEE journals and conferences.

## ANDREAS SPANIAS

**Andreas Spanias** is Professor in the School of Electrical, Computer, and Energy Engineering at Arizona State University (ASU). He is also the director of the Sensor Signal and Information Processing (SenSIP) center and the founder of the SenSIP industry consortium (now and NSF I/UCRC site). His research interests are in the areas of adaptive signal processing, speech processing, and sensor systems. He and his student team developed the computer simulation software Java-DSP and its award winning iPhone/iPad and Android versions. He is the author of two text books: Audio Processing and Coding by Wiley and DSP; An Interactive Approach (2nd Ed.). He served as Associate Editor of the IEEE Transactions on Signal Processing and as General Co-chair of IEEE ICASSP-99. He also served as the IEEE Signal Processing Vice-President for Conferences. Andreas Spanias is co-recipient of the 2002 IEEE Donald G. Fink paper prize award and was elected Fellow of the IEEE in 2003. He served as a Distinguished lecturer for the IEEE Signal processing society in 2004. He is a series editor for the Morgan & Claypool lecture series on algorithms and software.

Printed in the United States
by Baker & Taylor Publisher Services